# Carotenoid Cleavage Products

ACS SYMPOSIUM SERIES **1134**

# Carotenoid Cleavage Products

**Peter Winterhalter**, Editor
*Institute of Food Chemistry*
*Technische Universität Braunschweig*
*Braunschweig, Germany*

**Susan E. Ebeler**, Editor
*Department of Viticulture and Enology*
*University of California Davis Davis, California*

**Sponsored by the**
**ACS Division of Agriculture and Food Chemistry, Inc.**

American Chemical Society, Washington, DC

Distributed in print by Oxford University Press

**Library of Congress Cataloging-in-Publication Data**

Carotenoid cleavage products / Peter Winterhalter, Susan E. Ebeler, editors ; sponsored by the ACS Division of Agriculture and Food Chemistry, Inc.
pages cm. -- (ACS symposium series ; 1134)
Includes bibliographical references and index.
ISBN 978-0-8412-2778-1 (alkaline paper) 1. Carotenoids--Analysis--Congresses.
2. Carotenoids--Metabolism--Congresses. 3. Metabolites--Congresses.
4. Oxygenases--Congresses. 5. Enzymes--Congresses. 6. Food--Analysis--Congresses.
I. Winterhalter, Peter, 1957- II. Ebeler, Susan E., 1961- III. American Chemical Society. Division of Agricultural and Food Chemistry.
QK898.C34C36 2013
571.8′92--dc23

2013022427

The paper used in this publication meets the minimum requirements of American National Standard for Information Sciences—Permanence of Paper for Printed Library Materials, ANSI Z39.48n1984.



Distributed in print by Oxford University Press



PRINTED IN THE UNITED STATES OF AMERICA

# Foreword

The ACS Symposium Series was first published in 1974 to provide a mechanism for publishing symposia quickly in book form. The purpose of the series is to publish timely, comprehensive books developed from the ACS sponsored symposia based on current scientific research. Occasionally, books are developed from symposia sponsored by other organizations when the topic is of keen interest to the chemistry audience.

Before agreeing to publish a book, the proposed table of contents is reviewed for appropriate and comprehensive coverage and for interest to the audience. Some papers may be excluded to better focus the book; others may be added to provide comprehensiveness. When appropriate, overview or introductory chapters are added. Drafts of chapters are peer-reviewed prior to final acceptance or rejection, and manuscripts are prepared in camera-ready format.

As a rule, only original research papers and original review papers are included in the volumes. Verbatim reproductions of previous published papers are not accepted.

**ACS Books Department**

# Contents

## Biotechnology and Processing Effects



## Indexes

# Preface

## Carotenoid Cleavage Products

The importance of carotenoid cleavage reactions in plants and animals has been well-known since the early 1900s. These early studies demonstrated that β -carotene was degraded to form Vitamin A, which is important for human health and plays a critical role in vision. However, the actual biochemical mechanisms involved, and the enzyme responsible for β-carotene cleavage *in vivo*, were not identified until the mid- to late-1990s (Wolf 1995; Woggon 2000, von Lintig and Vogt) (*1–3*). Also in the late 1990's, similar carotenoid cleavage enzymes were identified in plants and found to be important in the production of plant hormones (Schwartz et al., 1997) (*4*) and aroma compounds.

These enzymatic cleavage reactions had only recently been discovered at the time of the last ACS symposium on Carotenoid Derived Aroma Compounds (Winterhalter and Rouseff, 2001) (*5*) and only three papers on enzymatic production of apo-carotenoids were presented. Since that time, the field has exploded and the importance of carotenoid cleavage enzymes in biological reactions has been well-established. The current ACS Symposium on Carotenoid Cleavage Products was designed to highlight these recent discoveries, focusing on the genetic and molecular biology of carotenoid cleavage enzymes, the importance of apo-carotenoids in flavor and aroma of fruits, vegetables, and wines, and the increasing interest in biotechnological aspects of apo-carotenoid production.

Following the Introduction (Chapter 1), the first section of this book (Chapters 2, 3, and 4) explores the general mechanisms of bio-oxidative carotenoid cleavage reactions, characterization of the genes involved in these cleavage reactions and the implications for biological activity of the apocarotenoids in plants and mammalian systems.

In the next section, (Chapters 5, 6, 7, and 8), the production of aroma-active compounds in spices, fruits, vegetables, and other plant systems is discussed. The potential biological activity as plant growth regulators is also considered.

Carotenoid cleavage products contribute significantly to the aroma of many varieties of grapes and wines, as discussed in Chapters 9, 10, 11, and 12. The impact of viticultural and wine-making practices as well as storage conditions are discussed. Many of the initial cleavage projects undergo modifications such as glycosylation and formation of acetylenic or allenic intermediates and the implications that these modifications have on grape and wine flavor are reviewed.

Finally, in Chapters 13, 14, 15, 16, 17, and 18, the biotechnological production of carotenoid cleavage products is discussed. The role that processing variables play in impacting carotenoids and subsequent cleavage products in a variety of food matrices, including paprika and eggs, are reviewed. We hope that this proceedings will be a valuable reference to food scientists, biochemists, and analytical chemists who are at the forefront of understanding the chemistry, analysis, and bioactivity of carotenoids and their cleavage products. We thank the contributors, the reviewers, and the following organizations (listed alphabetically) that financially contributed to making the ACS symposium possible: BREKO; Constellation Wines US, Inc.; DAAD; E&J Gallo Winery; ETS Laboratories; ifri Schuhmann; Mars, Inc.; Phytolab; and Symrise.

## References


1. Wolf, D. The enzymatic cleavage of β-carotene: Still controversial. *Nutr. Rev.* **1995**, *53*, 134–137.
2. Woggon, W.-D. The central cleavage of β,β-carotene-A supramolecular mimic of enzymatic catalysis. *Chimia* **2000**, *54*, 564–568.
3. von Lintig, J.; Vogt, K. J. Filling the gap in vitamin A research: Molecular identification of an enzyme cleaving ß-carotene to retinal. *Biol. Chem.* **2000**, *275*, 11915–11920.
4. Schwartz, S. H.; Tan, B. C.; Gage, D. A.; Zeevaart, J. A. D.; McCarty, D. R. Specific oxidative cleavage of carotenoids by VP14 of maize. *Science* **1997**, *276*, 1872–1874.
5. Winterhalter, P.; Rouseff, R. L. *Carotenoid-Derived Aroma Compounds*; ACS Symposium Series 802; American Chemical Society: Washington, DC, 2002.



**Peter Winterhalter**
Institute of Food Chemistry
Technische Universität Braunschweig
Schleinitzstraβe 20, DE-38106 Braunschweig, Germany

**Susan E. Ebeler**
Department of Viticulture and Enology
University of California, Davis
One Shields Avenue, Davis, CA 95616

# Bio-oxidative Cleavage

# Chapter 1

# Carotenoid Cleavage Products: An Introduction

**Peter Winterhalter*[,1] and Susan E. Ebeler[2]**

**[1]Institut für Lebensmittelchemie, Technische Universität Braunschweig, Schleinitzstraße 20, 38106 Braunschweig, Germany**
**[2]Department of Viticulture and Enology, University of California, Davis, One Shields Avenue, Davis, California 95616, United States**
***E-mail: p.winterhalter@tu-bs.de.**

During the last decade there has been tremendous progress in the area of bio-oxidative carotenoid cleavage and the resulting cleavage products. This introductory chapter will briefly describe the major classes of carotenoid metabolites as well as the major achievements in this field, i.e. the discovery of specific carotenoid cleavage enzymes and novel plant hormones, as well as several unexpected new functions of carotenoid metabolites.

The last ACS symposium on carotenoid metabolites in the year 2001 has had a strong focus on volatile carotenoid cleavage products. These compounds are widespread in the plant kingdom and constitute important ingredients for the flavor and perfume industry (*1*). Although the evidence for a formation *via* bio-oxidative carotenoid cleavage was very strong at that time, the responsible cleavage enzymes were still not known. Only in recent years the so-called carotenoid cleavage dioxygenases (CCDs) have been identified and a new insight into carotenoid metabolism has been obtained. The discovery of CCDs together with the detection of a novel group of plant hormones (strigolactones) lead to an enormous interest in carotenoid metabolism and the number of publications covering carotenoid cleavage grew almost exponentially. In view of these developments this symposium proceedings will try to briefly highlight recent discoveries in this rapidly growing research field.

## Carotenoid Cleavage Enzymes

Two different classes of carotenoid cleavage enzymes are known to be present in plants. The first class represents the so-called 9-cis-epoxycarotenoid dioxygenases (NCEDs) responsible for abscisic acid formation from neoxanthin and violaxanthin through cleavage of the carotenoid chain at the 11,12-position (*2*). The second class of cleavage enzymes are the so-called carotenoid cleavage dioxygenases (CCDs). The first CCD characterized on a molecular level was from *Arabidopsis thaliana* and labeled as *AtCCD1* (*3*). When expressed in *E. coli*, *AtCCD1* was found to cleave several carotenoids in the 9,10 and 9′,10′-position giving rise to a $C_{14}$-dialdehyde and two $C_{13}$-carotenoid endgroups. CCD1 is a non-heme enzyme, requiring only $Fe^{2+}$ as a cofactor. Convincing evidence for a dioxygenase mechanism has been obtained from labeling experiments (*4*). In total, the family of carotenoid cleavage enzymes in *A. thaliana* consists of 9 members (CCD 1, 4, 7, 8 and NCED 2, 3, 5, 6, and 9) (*5*, *6*). For an overview cf. refs. (*7–9*).

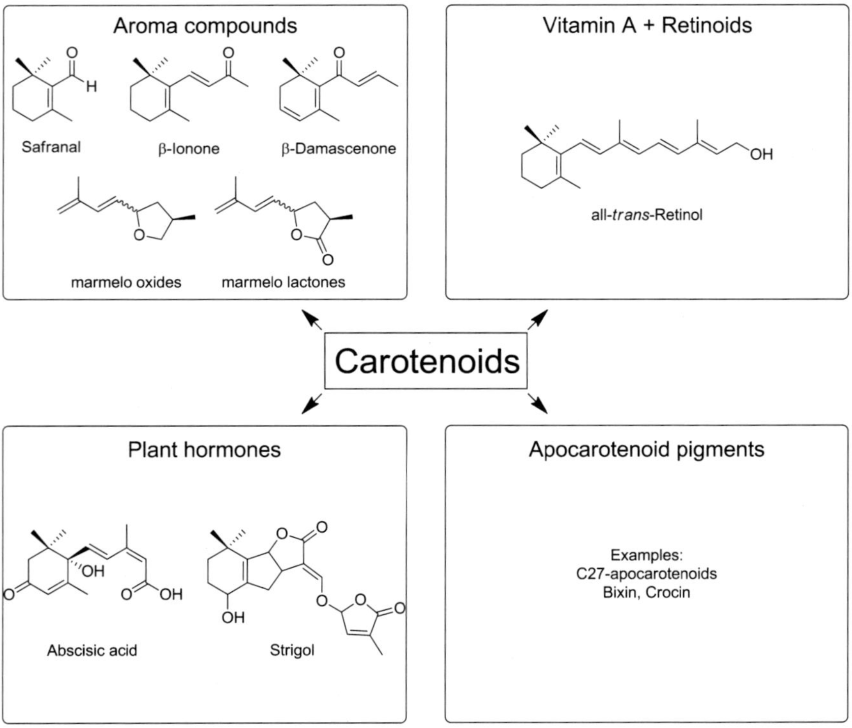


*Figure 1. Examples for carotenoid-derived metabolites.*

## Carotenoid Cleavage Products

Plant carotenoids are not only essential for photosynthesis and photoprotection (*10*) they also give rise to the formation of numerous biologically active cleavage products which *inter alia* include aroma compounds, vitamins, phytohormones, and apocarotenoid pigments (Figure 1).

### Aroma Compounds

A comprehensive overview on the occurrence and formation of carotenoid-derived aroma compounds from non-volatile precursors can be found in the ACS Symposium Volume 802 (*1*). Some structures of important volatiles are outlined in Figure 1. They include *inter alia* the $C_{10}$-compound safranal and the $C_{13}$-apocarotenoids ß-ionone and ß-damascenone which are obtained by cleavage of the carotenoid chain in the 7,8/7′,8′- and 9,10/9′,10′-position, respectively.

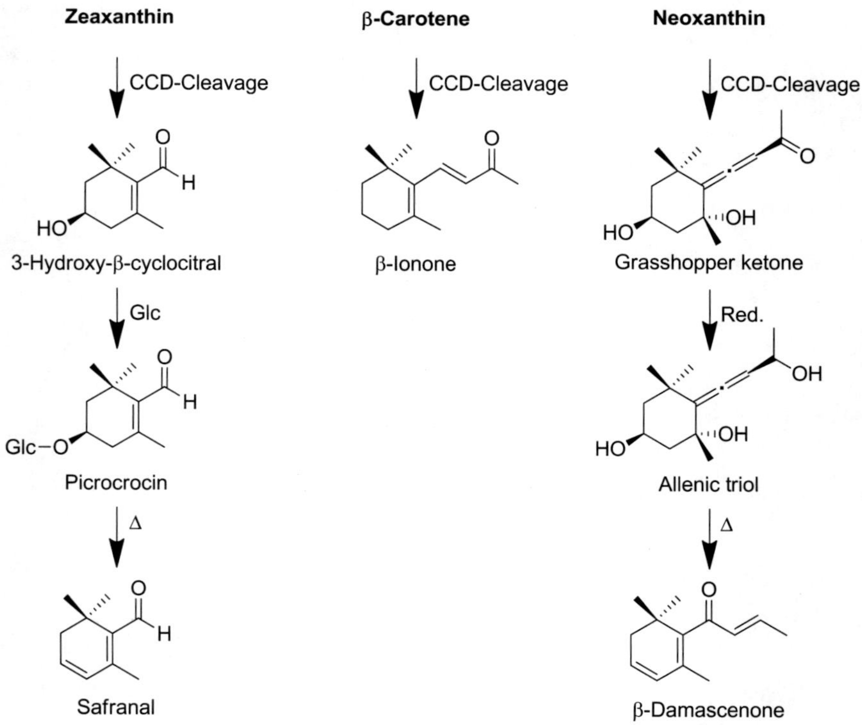


*Figure 2. Formation of key flavor compounds from the respective parent carotenoid.*

In many cases, a three-step mechanism is required for the formation of the aroma compounds in food: (i) CCD cleavage of the parent carotenoid giving rise to a primary cleavage product, (ii) enzymatic transformation (e.g., reduction and glycosylation) in plant tissues, and (iii) acid-catalyzed transformation during processing of food. Examples are shown in Figure 2. Whereas most of the aroma active apocarotenoids are derived from the carotenoid endgroup, a limited number is also obtained from the central portion of the carotenoid chain, such as, e.g., the odoriferous marmelo oxides and marmelo lactones (Figure 1).

With regard to CCD cleavage *in planta* a step-wise cleavage of the carotenoid has been suggested. Because of its localization in the cytosol, CCD1 is discussed as being mainly responsible for the cleavage of $C_{27}$-apocarotenoids. The latter are most likely generated by the action of a plastidial CCD (possible candidates CCD4 or CCD7) which have direct access to the intact carotenoids in the plastids and the so-obtained $C_{27}$-fragment is then expected to be translocated into the cytosol (*8*, *9*). This hypothesis is still under evaluation.

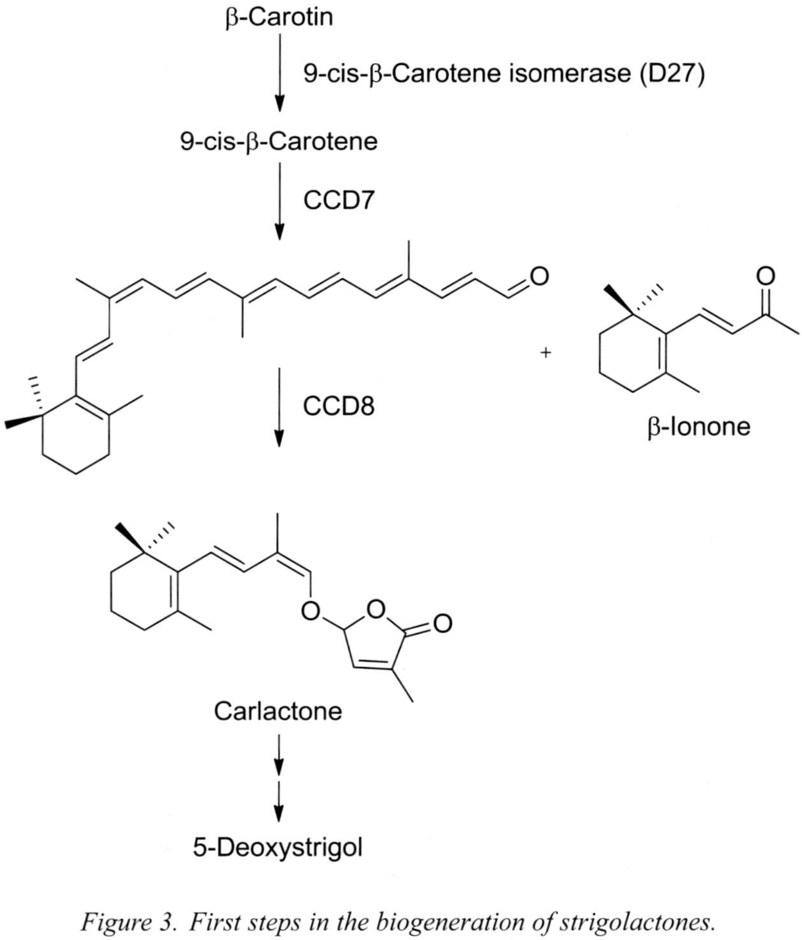


*Figure 3. First steps in the biogeneration of strigolactones.*

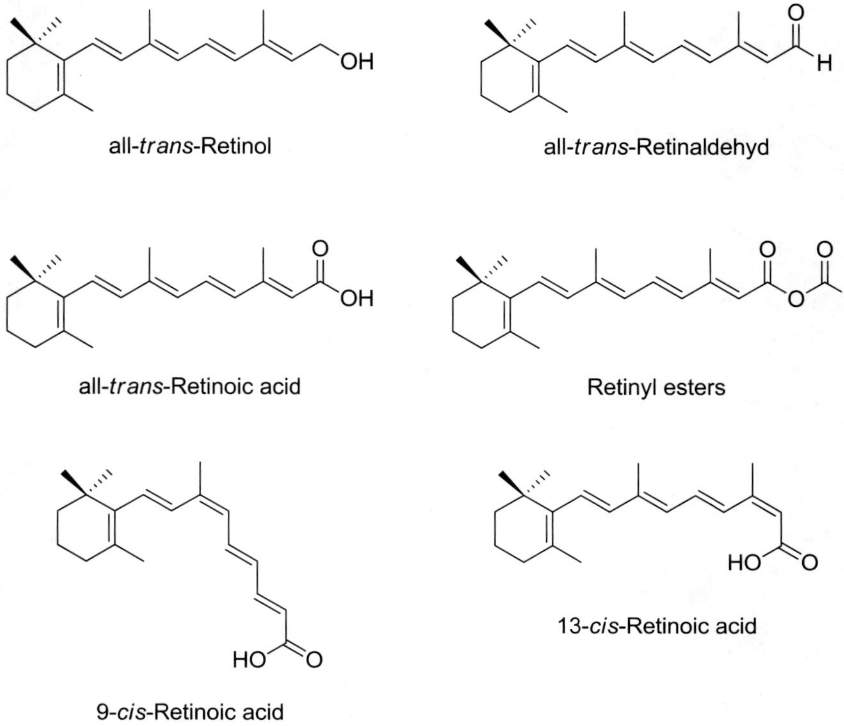


*Figure 4. Structure of retinoids.*

## Plant Hormones

Apart from the thoroughly studied plant hormone S-(+)-abscisic acid (*11*) that is formed by NCED-cleavage of 9-*cis*-violaxanthin and 9-*cis*-neoxanthin (*2*), the pathway of formation and function of a novel group of carotenoid-derived plant hormones, i.e. strigolactones, has only recently been elucidated (*12*, *13*). Strigolactones are known since the 1960s as allelochemicals that are secreted from the roots to the rhizosphere. To date, 15 different members have been structurally characterized. Strigolactones were first found to have crucial functions in induction of parasitic plant germination (e.g., witchweed) as well as in establishing the symbiosis with arbuscular mycorrhizal (AM) fungi. In 2008, strigolactones have been identified as plant hormones being responsible for shoot branching. Plants obviously regulate production of strigolactones in response to changes in nutrient supply. Under nutrient starvation, plants produce more strigolactones in order to minimize shoot branching and to promote symbiosis with AM fungi.

For the biogeneration of strigolactones the pathway outlined in Figure 3 has been elucidated in the year 2012 (*14*). After isomerization of all-*trans*-ß-carotene to the 9-*cis*-isomer by the isomerase D27, action of CCD7 gives rise to the formation of 9-cis-ß-apo-10′-carotenal, the latter being directly converted into carlactone by CCD8. Conversion of carlactone into the likely precursor of all other strigolactones, i.e., 5-deoxystrigol, is still under active investigation.

### Retinoids

Retinoids are a family of signaling molecules that are related to vitamin A. The structures of retinoids are outlined in Figure 4. Retinoids are derived from vitamin A which is formed by central cleavage of ß-carotene. Molecular identification of an enzyme cleaving ß-carotene to retinal has first been described in the year 2000 (*15*). This ß-carotene dioxygenase has been obtained from *Drosophila melanogaster* by expressing it into an *E. coli* strain. Whereas vitamin A is vital for vision and immune function, conversion to retinoic acid opens an avenue to even more active compounds which exert multiple effects on embryonic development, cell proliferation, differentiation and apoptosis. Retinoids are thus used in the treatment of human cancers (*16*).

## New Functions of Carotenoid Metabolites

Investigation of the biological activity of carotenoid cleavage products is a rapidly growing research field which will continue to produce novel and sometimes also unexpected results. In the case of volatile carotenoid cleavage products a screening of aroma compounds being present in apple fruit revealed a cancer chemopreventive potential of members of the damascone group, such as ß-damascenone, 3-HO-ß-damascone, and related substances (*17*). Other $C_{13}$-oxygenated apocarotenoids were identified as allelochemicals (*18*, *19*) and even the plant hormone abscisic acid has recently been found to possess an anti-inflammatory activity in mouse models (*20*). In view of this, carotenoid metabolites will continue to be a hot research topic in the future.

## References


1. *Carotenoid-derived Aroma Compounds*; Winterhalter, P., Rouseff, R. L., Eds.; ACS Symposium Series 802; American Chemical Society: Washington, DC, 2002.
2. Schwartz, S. H.; Tan, B. C.; Gage, D. A.; Zeevaart, J. A.; McCarty, D. R. *Science (Washington, DC)* **1997**, *276*, 1872–1874.
3. Schwartz, S. H.; Qin, X. Q.; Zeevaart, J. A. *J. Biol. Chem.* **2001**, *276*, 25208–25211.
4. Schmidt, H.; Kurtzer, R.; Eisenreich, W.; Schwab, W. *J. Biol. Chem.* **2006**, *281*, 9845–9851.
5. Bouvier, F.; Isner, J. C.; Dogbo, O.; Camara, B. *Trends Plant Sci.* **2005**, *10*, 187–194.

6. Auldridge, M. E.; McCarty, D. R.; Klee, H. J. *Curr. Opin. Plant Biol.* **2006**, *9*, 315–321.
7. Fleischmann, P.; Zorn, H. In *Carotenoids, Vol. 4: Natural Functions*; Britton, G., Liaaen-Jensen, S., Pfander, H. P., Eds.; Birkhäuser: Basel, 2008; pp 341–366.
8. Walter, M. H.; Floss, D. S.; Strack, D. *Planta* **2010**, *232*, 1–17.
9. Walter, M. H.; Strack, D. *Nat. Prod. Rep.* **2011**, *28*, 663–692.
10. Cazzonelli, C. I. *Funct. Plant Biol.* **2011**, *38*, 833–847.
11. Cutler, S. R.; Rodriguez, P. L.; Finkelstein, R. R.; Abrams, S. R. *Annu. Rev. Plant Biol.* **2010**, *61*, 651–67.
12. Ruyter-Spira, C.; Al-Babili, S.; van der Krol, S.; Bouwmester, H. *Trends Plant Sci.* **2013**, *18*, 72–83.
13. Seto, Y.; Kameoka, H.; Yamaguchi, S.; Kyozuka, J. *Plant Cell Physiol.* **2012**, *53*, 1843–1853.
14. Alder, A.; Jamil, M.; Marzorati, M.; Bruno, M.; Vermathen, M.; Bigler, P.; Ghisla, S.; Bouwmeester, H.; Beyer, P.; Al-Babili, S. *Science (Washington, DC)* **2012**, *335*, 1348–1351.
15. von Lintig, J.; Vogt, K. *J. Biol. Chem.* **2000**, *275*, 11915–11920.
16. Tang, X.-H.; Gudas, L. J. *Annu. Rev. Pathol.: Mech. Dis.* **2011**, *6*, 345–364.
17. Gerhäuser, C.; Klimo, K.; Hümmer, W.; Hölzer, J.; Petermann, A.; Garreta-Rufas, A.; Böhmer, F. D.; Schreier, P. *Mol. Nutr. Food Res.* **2009**, *53*, 1237–1244.
18. Dietz, H.; Winterhalter, P. *Phytochemistry* **1996**, *42*, 1005–1010.
19. Macias, F. A.; López, A.; Varela, R. M.; Torres, A.; Molinillo, J. M. G. *Phytochemistry* **2004**, *65*, 3057–3063.
20. Bassaganya-Riera, J.; Guri, A. J.; Lu, P.; Climent, M.; Carbo, A.; Sobral, B. W.; Horne, W. T.; Lewis, S. N.; Bevan, D. R.; Hontecillas, R. *J. Biol. Chem.* **2011**, *286*, 2504–2516.

## Chapter 2

# Carotenoid Cleavage Dioxygenase Genes from Fruit

**Wilfried Schwab,*,[1] Fong-Chin Huang,[1] and Péter Molnár[2]**

**[1]Biotechnology of Natural Products, Technische Universität München, Liesel-Beckmann-Strasse 1, D-85354 Freising, Germany**
**[2]University of Pécs, Medical School Department of Pharmacognosy, H-7624 Pécs, Rókus u. 2, Hungary**
***E-mail: schwab@wzw.tum.de.**

The oxidative cleavage of carotenoids and xanthophylls yields apocarotenoids and is catalyzed by a family of carotenoid cleavage dioxygenases (CCDs). Numerous CCDs have been cloned and functionally characterized since the discovery of the first carotenase gene *VP14* in 1997. In plants, *CCDs* are expressed in different tissues such as roots, shoots, leaves, flowers and fruits where they are implicated in various processes including plant growth regulation, plant hormone formation and production of volatiles. Since apocarotenals are important secondary metabolites that contribute to the quality of fresh fruit homologs of *CCD* genes have been identified in tomato, grape, apple, mandarin, orange, lemon, melon, peach and strawberry. Here, *CCDs* are discussed according to their roles in fruit.

Carotenoids and xanthophylls having fewer than 40 carbons can result from loss of carbons within the chain (norisoprenoids) or loss of carbons from the end of the molecule (apocarotenals). These changes are caused by oxidative degradation of the carotenoid and xanthophyll scaffolds and subsequent modification in a process termed ‘oxidative remodeling’ (Figure 1). The degradation products have significant roles in developmental and environmental response signaling as some of them may act as phytohormones (abscisic acid (ABA), strigolactones) and vitamins (retinal) (*1*). They also make important contributions to flavor and nutritional quality of several types of foods such as fruits, tea, wine, and tobacco. Two well-known natural apocarotenoids, bixin and crocetin, have economic importance as pigments and aroma in foods (*2*). Apocarotenoids also act as visual or volatile signals to attract pollinating agents and are also important in plant defense mechanisms (*3*). Studies have shown that the loss of certain carotenoid cleavage products induces the development of axillary branches, indicating that apocarotenoids convey signals that regulate plant architecture (*4*). Besides, a wide range of apocarotenals are produced by oxidative reactions during food processing and are intermediates in the formation of even smaller molecules of significance in food color and flavor.

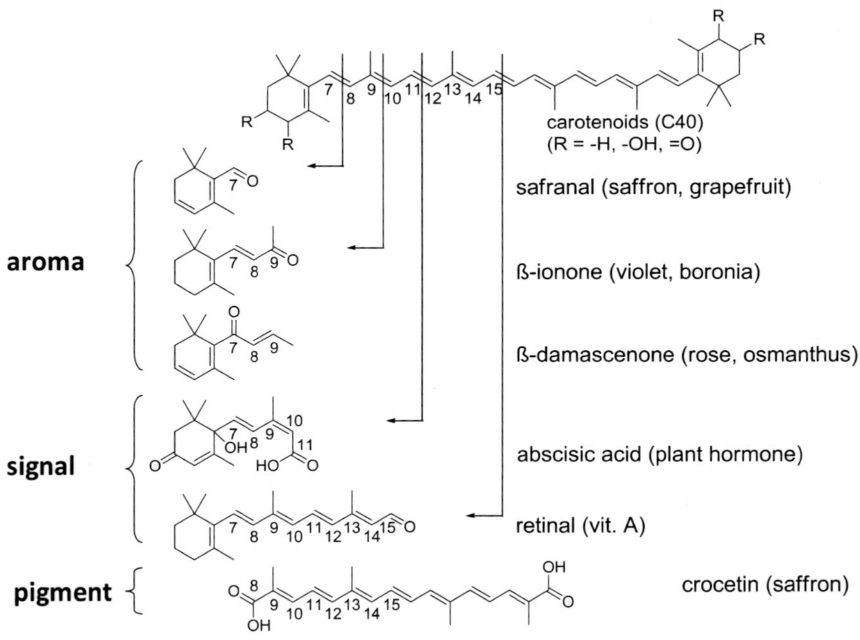


*Figure 1. Apocarotenals and norisoprenoids formed by regioselective oxidative cleavage of carotenoids and xanthophylls. Their occurrences and putative functions are indicated.*

In animals, the most important apocarotenoid is the $C_{20}$ compound retinal which plays a crucial role as the chromophore of rhodopsin in the vertebrate visual cycle (Figure 1). In plants and cyanobacteria, apocarotenoids are found in large amounts in the thylakoid membrane, where they act as accessory and photoprotective pigments.

# Carotenoid Cleavage Dioxygenases

In 1965, the enzymatic formation of retinal (vitamin A) from β-carotene was independently reported by two groups using crude enzyme fractions isolated from rat liver and intestines (*5*, *6*). It took another thirty years until an enzyme with carotenoid cleavage activity was isolated in pure form. A screen for viviparous maize seeds yielded a gene *VP14* encoding a protein that shows sequence homology to lignostilbene dioxygenase from *Pseudomonas paucimobilis* (*7*, *8*). Vivipary results from the lack of ABA, which had been postulated to be derived from the cleavage of a carotenoid in a reaction reminiscent of that catalyzed by lignostilbene dioxygenase. Enzymatic analysis of recombinant VP14 confirmed that it cleaves 9-cis-epoxycarotenoids such as 9-cis-violaxanthin and 9′-cis-neoxanthin, the first committed step in ABA biosynthesis. In the meantime, homology-based analysis using VP14 has identified many enzymes that are responsible for the synthesis of biologically important apocarotenoids (*9*). The proteins cleave the carotenoid backbone regiospecifically at different positions. One of the first of these was AtCCD1 from *Arabidopsis thaliana* (*10*). It cleaves a variety of carotenoids symmetrically at the 9,10 and 9′,10′ double bond to form a (di)aldehyde and one or two $C_{13}$ products (β-ionone), depending on the carotenoid substrate (Figure 1). Similar enzymes that cleave all-*trans* substrates have been identified in a number of other plants, and to distinguish them from the 9-*cis*-epoxy-carotenoid dioxygenases (NCED) within the carotenase family, they were called carotenoid cleavage dioxygenases (CCDs).

The carotenase family is ancient, with family members present in bacteria, plants and animals (*1*). Members of the carotenase family share a $Fe^{2+}$ ion in the catalytic site, four conserved histidines to coordinate iron binding and a conserved peptide sequence at their carboxyl terminius that minimally constitutes a signature sequence for the family (*11*, *12*). To date only two carotenase proteins have been crystallized and characterized in detail, namely Diox1 from the cyanobacterium *Synechocystis* sp. PCC6803 (*11*) and NCED1 (VP14) from maize (*12*). In both enzymes, the iron is embedded in a seven-bladed β-propeller chain arrangement. Coordination of a putative dioxygen molecule to the catalytic centre and the existence of a dioxocyclobutanyl (dioxetane) intermediate is the most likely scenario deduced from calculations of density maps and mechanistic considerations in the case of VP14. Unfortunately, attempts to co-crystallize VP14 with substrates failed (*12*).

A family of nine proteins, whose members are homologous to VP14, has been identified in the *Arabidopsis thaliana* genome and most of the members have been functionally characterized (*1*). Five of the *A. thaliana* proteins, like VP14, cleave 9-*cis*-epoxycarotenoids and are therefore called NCED2, 3, 5, 6, and 9. The remaining four are unlikely to be NCEDs and are instead termed CCD1, 4, 7, and 8 (Figure 2).

Screening of the published genome database of grape (*Vitis vinifera, (13)*), apple (*Malus domestica, (14)*) and strawberry (*Fragaria vesca, (15)*) yields sequences that show high similarity to NCEDs and CCDs from *A. thaliana* and provides evidence for a new subclass of CCDs in strawberry and grapes (Figure

2). None of these putative carotenase genes from grape, apple, and strawberry have been characterized except for *FaCCD1* (gene 26820) from *Fragaria x ananassa* (*16*) and *VvCCD1* (Vv1032110001) from *Vitis vinifera* (*17*).

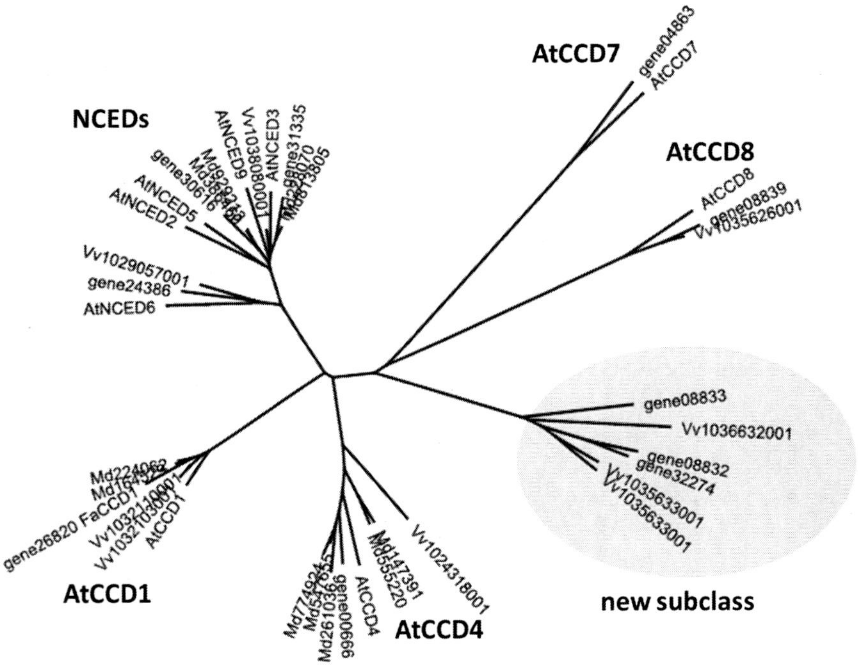


*Figure 2. Phylogenetic tree of CCDs and NCEDs from Arabidopsis thaliana (At), putative CCDs and NCEDs from grape (Vitis vinifera, Vv), apple (Malus domestica, Md) and strawberry (Fragaria veca, gene). Plant carotenases are grouped by substrate preference (e.g. NCEDS) and cleavage mechanism (regioselectivity). A novel subclass of carotenases was detected which does not contain a sequence from A. thaliana.*

## Reaction Mechanism of Carotenases

Most putative carotenoid cleavage enzymes have been termed dioxygenases and are currently classified as EC 1.13.11 (oxidoreductases; acting on single donors with incorporation of molecular oxygen; with incorporation of two atoms of oxygen). However, two reaction mechanisms have been postulated for CCDs (*9*). In the first step of the two mechanisms the oxygen (molecular oxygen dissolved in the aqueous environment) binds to the $Fe^{2+}$ and the substrate in the active site, creating a peroxide-intermediate. Afterwards, a dioxygenase mechanism oxidizes the double bond simultaneously on both carbons forming two carbonyls via a dioxetane intermediate. The dioxygenase mechanism has been experimentally confirmed for AtCCD1 with the help of isotopically labeled molecular oxygen (*18*). Surprisingly, the central cleavage of β-carotene in eukaryotes has been described to follow a mono-oxygenase mechanism, e.g. for chicken β-carotene-15,15′-oxygenase (*19*), whereas a bacterial Blh-protein (β-carotene-15,15′-oxygenase) is reported to use only the dioxygenase mechanism (*20*). The "carotenoid cleavage mono-oxygenases" are classified under EC 1.14.99 (oxidoreductases; acting on paired donors, with incorporation or reduction of molecular oxygen; miscellaneous).

## Carotenases in Fruit

In plants, *CCDs* are expressed in different tissues such as roots, shoots, leaves, flowers and fruits where they are implicated in various processes including plant growth regulation, plant hormone formation and production of volatiles (*1*). *CCD1*, *CCD4* and *NCEDs* are the major carotenases found in fruits and have been functionally characterized in different species (Table 1).

## CCD1

CCD1s are the most frequently investigated and therefore best-characterized group of CCDs in plants and fungi. They do not contain a signal peptide and are probably located in the cytosol whereas all other plant CCDs are transported into plastids. CCD1s have been shown to cleave a large spectrum of cyclic and non-cyclic carotenoids and apocarotenoids at the 9,10 and 9′,10′ double bonds *in vitro* (*16*, *17*, *21–23*). CCD1s from tomato and rose are promiscuous and cleave their substrates also at the 5,6 and 5′,6′ double bonds (*21*, *32*). *In vivo* studies have shown that CCD1 is involved in the formation of volatiles such as β-ionone, geranylacetone, pseudoionone and 6-methyl-5-hepten-2-one (*21*, *32*). Both $C_8$ and $C_{13}$ apocarotenoid derivatives are potent scent, flavor and aroma compounds. Therefore, it is assumed that CCD1 plays an important role in fruit flavor formation. Recent experiments with interference RNA (RNAi) suggest that the *in vivo* substrate is a $C_{27}$ apocarotenoid – at least in *Medicago truncatula* (*33*). Upon silencing of tomato CCD1, fruit emissions of β-ionone and geranyl acetone decrease, demonstrating a link between CCD1 and apocarotenoid production *in vivo* (*21*). Although CCD1 expression is elevated during fruit development, it does not precisely mirror the emission of apocarotenoid volatiles (*16*, *17*, *21*). The time period between CCD1 gene expression and apocarotenoid production is greater then expected, probably because of differences in subcellular localization and the consequent availability of substrates (*3*).

**Table 1. Functionally Characterized Carotenoid Cleavage Dioxygenase Genes from Fruit Crop**

| *gene* | *fruit* | *reference* |
|---|---|---|
| **CCD1** | tomato (*Lycopersicon esculentum*) | (*21*) |
| | melon (*Cucumis melo*) | (*22*) |
| | grape (*Vitis vinifera*) | (*17*) |
| | citrus (*C. limon, C. sinensis, C. unshiu*) | (*23*) |
| | strawberry (*Fragaria x ananassa*) | (*16*) |
| **CCD4** | apple (*Malus x domestica*) | (*24*) |
| | peach (*Prunus persica*) | (*25*) |
| **NCED** | strawberry (*Fragaria x ananassa*) | (*26*) |
| | citrus (*C. limon, C. sinensis, C. unshiu*) | (*23, 27*) |
| | grape (*Vitis vinifera*) | (*28, 29*) |
| | peach (*Prunus persica*) | (*29*) |
| | persimmon (*Diospyros kaki*) | (*30*) |
| | avocado (*Persea americana*) | (*31*) |

Up to now, at least twelve reports on CCD1 genes and enzymes have been published covering many flowers and fruit of commercial importance. CCD1 is considered to be the prime candidate for enzymatic $C_{13}$ volatile apocarotenoid biogenesis.

## CCD4

Heterologous expression of the *MdCCD4* gene from apple reduced accumulation of β-carotene and zeaxanthin only slightly in *Escherichia coli* strains engineered for the accumulation of these carotenoids/xanthophylls (*24*). As for volatile cleavage products produced in these strains, only β-ionone was detected by solid-phase-micro-extraction GC-MS. The result implies 9,10 and/or 9′,10′ cleavage activity for MdCCD4, which is in agreement with other studies on CCD4 (*25*, *34*). In one of these studies silencing the *CCD4a* gene by RNAi resulted in a change of petal color from white to yellow, demonstrating that the accumulation of carotenoid in *Chrysanthemum* petals occurred only in the absence of CCD4a activity. The results implies carotenoid and/or xanthophylls cleavage activity of the CCD4a gene product (*34*). It seems that CCD1 and CCD4 enzymes cleave at the same positions within the carotenoid scaffold but are located in different compartments within plant cells.

## NCED

The carotenase class involved in plant ABA biosynthesis, called NCEDs is unique among the carotenases in that it accepts only *cis*-isomers of its substrates 9-*cis*-violaxanthin and 9′-*cis*-neoxanthin (*7*). *A. thaliana* produces five different NCEDs targeted to different locations inside the plastids due to their signal peptides (*1*). The primary $C_{15}$ product (xanthoxin) of NCED cleavage activity must subsequently be exported into the cytosol and further processed to ABA. Down-regulation of *NCED* genes in different plants cause severe ABA deficiency and wilty phenotypes (*1*).

ABA has been suggested to play also an important role in fruit development. Transient down-regulation of *NCED1* expression during strawberry fruit ripening resulted in a significant decrease in ABA levels and uncolored fruits (*26*). A similar uncolored phenotype was observed in transgenic RNA interference (RNAi) fruits, in which the expression of a putative ABA receptor gene was down-regulated. More importantly, the uncolored phenotype of the *FaNCED1*-down-regulated RNAi fruits could be rescued by exogenous ABA, These data provide evidence that ABA is a signal molecule that promotes strawberry ripening and emphasizes the significance of NCEDs in fruit ripening.

## Conclusion

The enzymatic formation of carotenoid and xanthophyll cleavage products and the genes involved have become a very active area of research. Elucidation of the 3-dimensional structures of prototype carotenase proteins has now shed light on the mechanisms of the enzymatic oxidative cleavage of poly-unsaturated compounds and show how CCDs associate with the membrane as a way of gaining access to its membrane soluble substrate. New carotenases and biological functions of apocarotenoids are continuously being identified. The basis for comparative genomics is broadening, and this approach will become an important tool to clarify evolutionary relationships.

## References


1. Walter, M. H.; Strack, D. *Nat. Prod. Rep.* **2011**, *28*, 663–692.
2. Bouvier, F.; Dogbo, O.; Camara, B. *Science (Washington, D.C.)* **2003**, *300*, 2089–2091.
3. Auldridge, M. E.; McCarty, D R.; Klee, H. J. *Curr. Opin. Plant Biol.* **2006**, *9*, 315–321.
4. Booker, J.; Auldridge, M.; Wills, S.; McCarty, D.; Klee, H.; Leyser, O. *Curr. Biol.* **2004**, *14*, 1232–1238.
5. Olson, J. A.; Hayaishi, O. *Proc. Natl. Acad. Sci. U.S.A.* **1965**, *54*, 1364–1370.
6. Goodman, D. S.; Huang, H. S. *Science (Washington, D.C.)* **1965**, *149*, 879–880.
7. Schwartz, S. H.; Tan, B. C.; Gage, D. A.; Zeevaart, J. A.; McCarty, D. R. *Science (Washington, D.C.)* **1997**, *276*, 1872–1874.

8. Tan, B. C.; Schwartz, S. H.; Zeevaart, J. A.; McCarty, D. R. *Proc. Natl. Acad. Sci. U.S.A.* **1997**, *94*, 12235–12240.
9. Kloer, D. P.; Schulz, G. E. *Cell Mol. Life Sci.* **2006**, *63*, 2291–2303.
10. Schwartz, S. H.; Qin, X.; Zeevaart, J. A. *J. Biol. Chem.* **2001**, *276*, 25208–25211.
11. Kloer, D. P.; Ruch, S.; Al-Babili, S.; Beyer, P.; Schulz, G. E. *Science (Washington, D.C.)* **2005**, *308*, 267–269.
12. Messing, S. A. J.; Gabelli, S. B.; Echeverria, I.; Vogel, J. T.; Guan, J. C.; Tan, B. C.; Klee, H. J.; McCarty, D. R.; Amzel, L. M. *Plant Cell* **2010**, *22*, 2970–2980.
13. Jaillon, O.; Aury, J. M.; Noel, B.; Policriti, A.; Clepet, C.; Casagrande, A.; Choisne, N.; Aubourg, S.; Vitulo, N.; Jubin, C.; Vezzi, A.; Legeai, F.; Hugueney, P.; Dasilva, C.; Horner, D.; Mica, E.; Jublot, D.; Poulain, J.; Bruyère, C.; Billault, A.; Segurens, B.; Gouyvenoux, M.; Ugarte, E.; Cattonaro, F.; Anthouard, V.; Vico, V.; Del Fabbro, C.; Alaux, M.; Di Gaspero, G.; Dumas, V.; Felice, N.; Paillard, S.; Juman, I.; Moroldo, M.; Scalabrin, S.; Canaguier, A.; Le Clainche, I.; Malacrida, G.; Durand, E.; Pesole, G.; Laucou, V.; Chatelet, P.; Merdinoglu, D.; Delledonne, M.; Pezzotti, M.; Lecharny, A.; Scarpelli, C.; Artiguenave, F.; Pè, M. E.; Valle, G.; Morgante, M.; Caboche, M.; Adam-Blondon, A. F.; Weissenbach, J.; Quétier, F.; Wincker, P. *Nature (London)* **2007**, *449*, 463–467.
14. Velasco, R.; Zharkikh, A.; Affourtit, J.; Dhingra, A.; Cestaro, A.; Kalyanaraman, A.; Fontana, P.; Bhatnagar, S. K.; Troggio, M.; Pruss, D.; Salvi, S.; Pindo, M.; Baldi, P.; Castelletti, S.; Cavaiuolo, M.; Coppola, G.; Costa, F.; Cova, V.; Dal Ri, A.; Goremykin, V.; Komjanc, M.; Longhi, S.; Magnago, P.; Malacarne, G.; Malnoy, M.; Micheletti, D.; Moretto, M.; Perazzolli, M.; Si-Ammour, A.; Vezzulli, S.; Zini, E.; Eldredge, G.; Fitzgerald, L. M.; Gutin, N.; Lanchbury, J.; Macalma, T.; Mitchell, J. T.; Reid, J.; Wardell, B.; Kodira, C.; Chen, Z.; Desany, B.; Niazi, F.; Palmer, M.; Koepke, T.; Jiwan, D.; Schaeffer, S.; Krishnan, V.; Wu, C.; Chu, V. T.; King, S. T.; Vick, J.; Tao, Q.; Mraz, A.; Stormo, A.; Stormo, K.; Bogden, R.; Ederle, D.; Stella, A.; Vecchietti, A.; Kater, M. M; Masiero, S.; Lasserre, P.; Lespinasse, Y.; Allan, A. C.; Bus, V.; Chagné, D.; Crowhurst, R. N.; Gleave, A. P.; Lavezzo, E.; Fawcett, J. A.; Proost, S.; Rouzé, P.; Sterck, L.; Toppo, S.; Lazzari, B.; Hellens, R. P.; Durel, C. E.; Gutin, A.; Bumgarner, R. E.; Gardiner, S. E.; Skolnick, M; Egholm, M.; Van de Peer, Y.; Salamini, F.; Viola, R. *Nat. Genet.* **2010**, *42*, 833–839.
15. Shulaev, V.; Sargent, D. J.; Crowhurst, R. N.; Mockler, T. C.; Folkerts, O.; Delcher, A. L.; Jaiswal, P.; Mockaitis, K.; Liston, A.; Mane, S. P.; Burns, P.; Davis, T. M.; Slovin, J. P.; Bassil, N.; Hellens, R. P.; Evans, C.; Harkins, T.; Kodira, C.; Desany, B.; Crasta, O. R.; Jensen, R. V.; Allan, A. C.; Michael, T. P.; Setubal, J. C.; Celton, J. M.; Rees, D. J. G.; Williams, K. P.; Holt, S. H.; Ruiz Rojas, J. J.; Chatterjee, M.; Liu, B.; Silva, H.; Meisel, L.; Adato, A.; Filichkin, S. A.; Troggio, M.; Viola, R.; Lynn Ashman, T.; Wang, H.; Dharmawardhana, P.; Elser, J.; Raja, R.; Priest, H. D.; Bryant, D. W.; Fox, S. E.; Givan, S. A.; Wilhelm, L. J.; Naithani, S.; Christoffels, A.; Salama, D.

Y.; Carter, J.; Lopez Girona, E.; Zdepski, A.; Wang, W.; Kerstetter, R. A.; Schwab, W.; Korban, S. S.; Davik, J.; Monfort, A.; Denoyes-Rothan, B.; Arus, P.; Mittler, R.; Flinn, B.; Aharoni, A.; Bennetzen, J. L.; Salzberg, S. L.; Dickerman, A. W.; Velasco, R.; Borodovsky, M.; Veilleux, R. E.; Folta, K. M. *Nat. Genet.* **2011**, *43*, 109–118.

16. García-Limones, C.; Schnäbele, K.; Blanco-Portales, R.; Luz Bellido, M.; Caballero, J. L.; Schwab, W.; Muñoz-Blanco, J. *J. Agric. Food Chem.* **2008**, *56*, 9277–9285.
17. Mathieu, S.; Terrier, N.; Procureur, J.; Bigey, F.; Günata, Z. *J. Exp. Bot.* **2005**, *56*, 2721–2731.
18. Schmidt, H.; Kurtzer, R.; Eisenreich, W.; Schwab, W. *J. Biol. Chem.* **2006**, *281*, 9845–9851.
19. Leuenberger, M. G.; Engeloch-Jarret, C.; Woggon, W. D. *Angew. Chem.* **2001**, *40*, 2613–2617.
20. Kim, Y. S.; Kim, N. H.; Yeom, S. J.; Kim, S. W.; Oh, D. K. *J. Biol. Chem.* **2009**, *284*, 15781–15793.
21. Simkin, A. J.; Schwartz, S. H.; Auldridge, M.; Taylor, M. G.; Klee, H. J. *Plant J.* **2004**, *40*, 882–892.
22. Ibdah, M.; Azulay, Y.; Portnoy, V.; Wasserman, B.; Bar, E.; Meir, A.; Burger, Y.; Hirschberg, J.; Schaffer, A. A.; Katzir, N.; Tadmor, Y.; Lewinsohn, E. *Phytochemistry* **2006**, *67*, 1579–1589.
23. Kato, M.; Matsumoto, H.; Ikoma, Y.; Okuda, H.; Yano, M. *J. Exp. Bot.* **2006**, *57*, 2153–2164.
24. Huang, F. C.; Molnár, P.; Schwab, W. *J. Exp. Bot.* **2009**, *60*, 3011–3022.
25. Brandi, F.; Bar, E.; Mourgues, F.; Horváth, G.; Turcsi, E.; Giuliano, G.; Liverani, A.; Tartarini, S.; Lewinsohn, E.; Rosati, C. *BMC Plant Biol.* **2011**, *11*, 24 doi:10.1186/1471-2229-11-24.
26. Jia, H. F.; Chai, Y. M.; Li, C. L.; Lu, D.; Luo, J. J.; Qin, L.; Shen, Y. Y. *Plant Physiol.* **2011**, *157*, 188–199.
27. Rodrigo, M. J.; Alquezar, B.; Zacarías, L. *J. Exp. Bot.* **2006**, *57*, 633–643.
28. Sun, L.; Zhang, M.; Ren, J.; Qi, J.; Zhang, G.; Leng, P. *BMC Plant Biol.* **2010**, *10*, 257 doi:10.1186/1471-2229-10-257.
29. Zhang, M.; Leng, P.; Zhang, G.; Li, X. *J. Plant Physiol.* **2009**, *166*, 1241–1252.
30. Leng, P.; Zhang, G. L.; Li, X. X.; Wang, L. H.; Zheng, Z. M. *Chin. Sci. Bull.* **2009**, *54*, 2830–2838.
31. Chernys, J. T.; Zeevaart, J. A. *Plant Physiol.* **2000**, *124*, 343–353.
32. Huang, F. C.; Horváth, G.; Molnár, P.; Turcsi, E.; Deli, J.; Schrader, J.; Sandmann, G.; Schmidt, H.; Schwab, W. *Phytochemistry* **2009**, *70*, 457–464.
33. Floss, D. S.; Walter, M. H. *Plant Signal. Behav.* **2009**, *4*, 172–175.
34. Ohmiya, A.; Kishimoto, S.; Aida, R.; Yoshioka, S.; Sumitomo, K. *Plant Physiol.* **2006**, *142*, 1193–1201.

## Chapter 3

# Involvement of CCD4 in Determining Petal Color

**A. Ohmiya***

**National Institute of Floricultural Science,
Tsukuba, Ibaraki 305-8519, Japan
*E-mail: ohmiya@affrc.go.jp.**

In chrysanthemums (*Chrysanthemum morifolium* Ramat.), white petal color is dominant over yellow and is postulated to arise from a single dominant gene that inhibits carotenoid formation. By differential screening, we identified a gene that is expressed specifically in white petals of chrysanthemum. The gene is highly homologous to *carotenoid cleavage dioxygenase 4* (*CCD4*) and was designated *CmCCD4a*. Suppression of *CmCCD4a* expression by RNA interference in white-flowered strains resulted in the formation of yellow petals. Flower color mutants with increased levels of carotenoids showed decreased levels of *CmCCD4a* expression. The results indicate that *CmCCD4a* is the single dominant gene encoding an enzyme that inhibits carotenoid accumulation in chrysanthemum petals by catalyzing degradation of carotenoids. Although CmCCD4a has been shown to catalyze the conversion of β-carotene to β-ionone *in vivo*, its catalytic product(s) *in planta* has not been identified. Involvement of CCD4 in controlling carotenoid content was also examined in petals of white-flowered *Ipomoea nil* and carnation (*Dianthus caryophyllus* L.). The results showed that low levels of carotenoids in petals of these plants were not caused by degradation but by low rates of biosynthesis.

## Introduction

Carotenoids are major plant pigments, and yellow flower colors of many plants are derived from carotenoids (*1*). Many other plants lack carotenoids in their petals, and instead express white, pink, purple, and blue petal colors. Furthermore, flowers have a wide range of carotenoid content, from little or none to large amounts even within the same plant species. However, not much is known about the mechanisms that underlie these differences.

In general, pigmented flowers are dominant over white flowers, and mutation usually occurs in the direction of pigmented to white flowers. However, in the case of chrysanthemum (*Chrysanthemum morifolium* Ramat.), petal color mutations occur from white to yellow (*2–5*). On the basis of an experiment in which white- and yellow-flowered chrysanthemums were crossed, Hattori (*6*) concluded that the white petal color is dominant over yellow and postulated a single dominant gene that inhibits carotenoid formation. The function of such a gene, however, was unknown.

To find the single dominant gene, a series of studies was conducted (*7–12*). In this paper, we show that a gene encoding carotenoid cleavage dioxygenase 4 (designated CmCCD4a) is the single dominant gene. However, rather than inhibiting carotenoid formation, the product of this gene catalyzes breakdown of carotenoids. This chapter also shows some examples of other species in which carotenoid content in petals is controlled not by degradation but by suppression of the flux through the carotenoid biosynthesis pathway.

## Cloning and Characterization of *CmCCD4a*

To identify the single dominant gene that causes inhibition of carotenoid accumulation in white petals of chrysanthemum, subtraction screening between white and yellow petals was performed (*7*). We obtained a gene that was expressed in a color-specific manner; the expression was very high in white petals and extremely low in yellow petals. The deduced amino acid sequence of the cloned gene (designated *CmCCD4a*) has high similarity to carotenoid cleavage dioxygenase 4 (Figure 1). Expression of *CmCCD4a* was strictly limited to petals and was extremely low in leaves, stems, and roots. The level of *CmCCD4a* transcripts in the yellow disc florets was also extremely low compared to the level in the white petals of ray florets.

To determine the role of CmCCD4a protein in the formation of petal color, we produced transgenic chrysanthemum plants with reduced expression of *CmCCD4a* by introducing an RNA interference construct of *CmCCD4a* into white-flowered chrysanthemum cultivars. The petal color of transformants was yellow due to increased levels of carotenoids (*7*, *8*). These results clearly show that white petals do synthesize carotenoids but the carotenoids are degraded by CmCCD4a. To confirm whether white petals synthesize carotenoids, we compared carotenogenic gene expression in white and yellow petals (*10*). Most of the carotenogenic genes are expressed as well in white petals as in yellow petals, indicating that both yellow and white petals synthesize carotenoids.

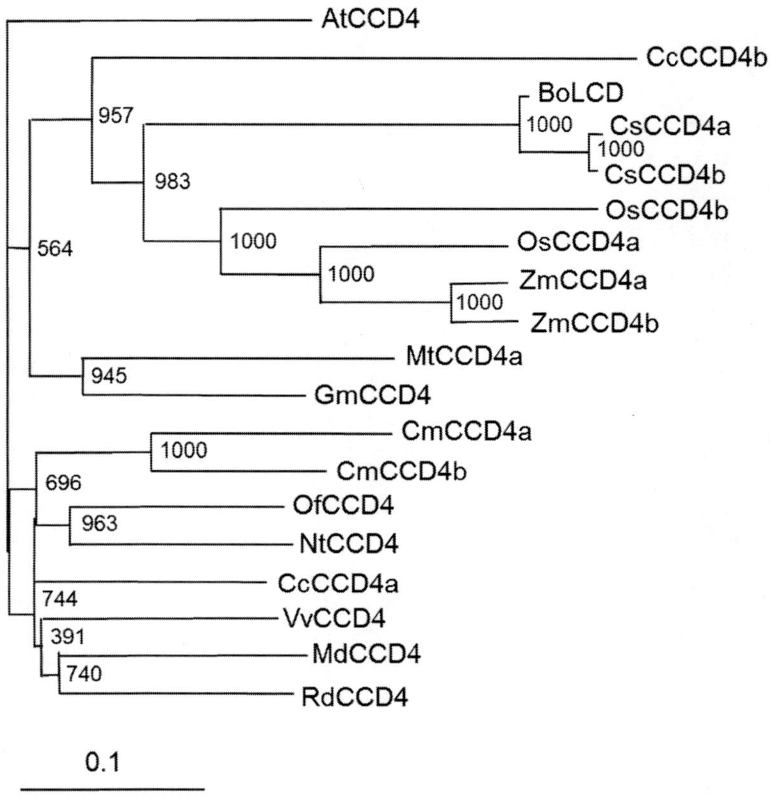


*Figure 1. ClustalW tree analysis of CCD4 in various plant species. Arabidopsis thaliana (AtCCD4, At4g19170), Bixa orellana (BoLCD, AJ489277), Citrus clementina (CcCCD4a, DQ309330; CcCCD4b, DQ309331), Chrysanthemum morifolium (CmCCD4a, AB247158; CmCCD4b, AB247160), Crocus sativus (CsCCD4a, EU523662; CsCCD4b, EU523663), Glycine max (NC_016088), Malus domestica (MdCCD4, EU327777), Medicago truncatula (MtCCD4a, MTR_5g025250), Nicotiana tabacum (NtCCD4, JF947192), Osmansus fragrans (OfCCD4, EU334434), Oryza sativa (OsCCD4a, Os02g47510; OsCCD4b, Os12g24800), Rosa x damascena (RdCCD4, EU334433), Vitis vinifera (vvi:100251100), Zea mays (ZmCCD4a, AC190588; ZmCCD4b, AC194862). Numbers at branch points indicate bootstrap values (1,000 replicates). The 0.1 scale bar means 0.1 nucleotide substitutions per site.*

CCD1 has been shown to contribute to the formation of apocarotenoid volatiles in the fruits and flowers of several plant species (*13–16*). We compared *CCD1* expression in white and yellow petals of chrysanthemum. The expression of *CCD1* was different among cultivars but there was no correlation between the expression of *CCD1* and carotenoid content (Figure 2). These results suggest that CCD1 is not directly involved in the regulation of carotenoid levels in petals.

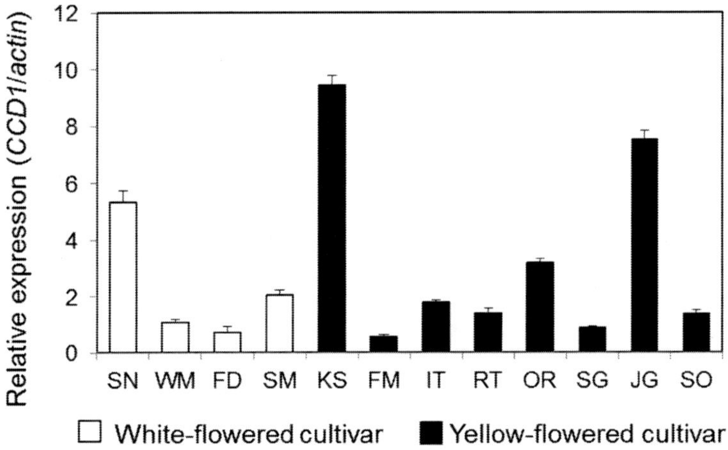


*Figure 2. Expression levels of CCD1 in petals of white- and yellow-flowered chrysanthemum cultivars. Expression levels were normalized against mRNA levels of actin; mean values (±SE) are shown (n=3). Petals from fully opened flowers were used for the analysis. SN, Seiko-no-Makoto; WM, White Marble; FD, Fiducia; SM, Sei-Marine; KS, Kurenai-Seiko; FM, Florida Marble; IT, Intole; RT, Ritz; OR, Olrina; SG, Seigetsu; JG, Jersey; SO, Sunny Orange.*

## Involvement of CCD4 in the Petal Color Mutation

Crossing experiments showed that *CmCCD4a* is a dominant gene that is homozygous in white-flowered chrysanthemums (*11*). These results made *CmCCD4a* a strong candidate for the single dominant gene that inhibits carotenoid accumulation as described above. To test this hypothesis, we analyzed the relationship between *CmCCD4a* expression and carotenoid content in petal color mutants.

By applying heavy-ion-beam radiation to the white-flowered chrysanthemum cultivar 'Jimba', we obtained mutants with pale yellow flowers (IB-1 lines) (*9*). IB-1 lines were reirradiated and mutants with much deeper yellow flowers were obtained (IB-2 lines). Increased carotenoid content in petals of the mutants was well correlated with decreased expression of *CmCCD4a*. *CmCCD4a* may represent a small gene family rather than a single gene, because chrysanthemum cultivars are hexaploid ($2n = 6x = 54$) (*17*, *18*). In fact, at least 6 homologs were found in the 'Jimba' genome; they have more than 92% sequence similarity (*9*). Four of them, *CmCCD4a-1*, *-2*, *-3*, and *-5*, were expressed in ray petals of 'Jimba'. Expression of *CmCCD4a-1, -2,-3, and -5* was strictly limited to petals and was extremely low in leaves, stems, and roots. *CmCCD4a-4* is a pseudogene with in-frame stop codons in the coding region. Petals of IB-1 lines express *CmCCD4a-2*, *-3*, and *-5* but lack *CmCCD4a-1* expression. Only *CmCCD4a-5* was expressed in IB-2 lines. The results suggest that the *CmCCD4a-1* was lost at the first irradiation, and *CmCCD4a-2* and *-3* were lost at the second irradiation.

The results of genomic PCR analysis using primers specific to each homolog support the hypothesis: a PCR product corresponding to *CmCCD4a-1* was not detected in IB-1 lines and those corresponding to *CmCCD4a-1*, *-2* and *-3* were not detected in IB-2 lines.

Loss of *CmCCD4* expression was also observed in the petal color mutants produced by bud sports (*12*). Petals of chrysanthemums are composed of superficial L1 and internal L2 cell layers (*3*). Each cell layer is derived from a different somatic cell, whose genome can mutate independently. Therefore, flowers that arise from bud sports are, in general, periclinal chimeras with genetically different cell layers. Many yellow-flowered bud sports of chrysanthemums are chimeras with carotenoids in only one cell layer (*19*). In case of the 'Marble' cultivars, a series of bud sports have various carotenoid contents in petals. Petals of 'White Marble' contain carotenoids in neither L1 nor L2, while those of 'Polished Marble' are chimeric and contain carotenoids only in L2 (*20*). Petals of 'Florida Marble' contain carotenoids in both L1 and L2 and have more total carotenoid than do those of 'Polished Marble' (Figure 3). *CmCCD4a-2* was expressed in petals of 'White Marble' and 'Polished Marble' but was not in those of 'Yellow Marble' (*11*). Because carotenoids accumulate in L2 of 'Polished Marble', *CmCCD4a-2* expression might be limited to L1. During the breeding process of 'Marble' cultivars, loss of *CmCCD4a-2* might have first occurred in L2 and then in L1. A stepwise decrease in the amount of *CmCCD4a* expression could result in a corresponding stepwise increase in the total carotenoid content in ray petals.

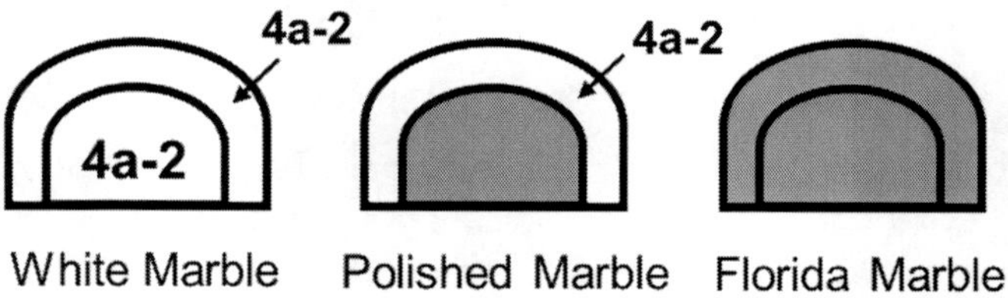


*Figure 3. Schematic diagrams of the chimeric pigment distribution in ray petals of chrysanthemum 'Marble' cultivars. Layer L1 covers layer L2. Carotenoid accumulation patterns (indicated by gray) are based on the results described by Shibata and Kawata (20). Expression patterns of CmCCD4a-2 (indicated by "4a-2") are based on the results described by Yoshioka et al (11).*

Dowrick and El-Bayoumi (*21*) showed that changes in chromosome number and chromosome fragmentation are sometimes responsible for changes in petal color of chrysanthemum. During the process of petal color mutation, loss of *CmCCD4a* gene(s) may occur as a result of loss of a chromosome, and cause a change in carotenoid accumulation. I therefore propose that *CmCCD4a* is the single dominant gene that causes inhibition of carotenoid accumulation; however, the results suggest that rather than inhibiting carotenoid formation, its product catalyzes carotenoid breakdown.

The number of *CmCCD4a* homologs functioning in chrysanthemum petals differ among cultivars, which may affect the frequency of petal color mutation. As described above, 'White Marble' has only *CmCCD4a-2* (*12*). Petal color mutation from white to yellow sometimes occurs in 'White Marble'. In contrast, 'Jimba' has at least 6 *CmCCD4a* homologs and 4 of them are expressed in petals (*9*). In such cultivars, loss of one or two *CmCCD4a* genes cannot lead to complete loss of CmCCD4a enzyme activity and petal color mutation from white to yellow may not occur. In fact, spontaneous mutation toward yellow petal color has rarely been found in 'Jimba'.

It is of great interest to know whether the mechanism of white petal color formation found in chrysanthemum also applicable to other plant species. *CCD4* orthologs have been isolated and characterized from many plant species (Figure 1) (*22–26*). A few of them were reported to be involved in controlling carotenoid content in flowers. In rose (*Rosa damascena*), 15 apocarotenoids have been identified in flowers, and the existence of a CCD that cleaves diverse carotenoid substrates has been postulated (*27*). Recently, orthologs of *CCD1* and *CCD4* were identified in rose, and their possible involvement in carotenoid degradation in flowers was demonstrated (*24*). In the stigma of crocus (*Crocus sativus*), *CCD1* and *CCD4* were highly expressed and considered to be involved in the formation of apocarotenoids in the stigma tissues (*23*). *CCD1* and *CCD4* were also shown to be expressed in crocus petals but the functions of these enzymes in petals have not been fully elucidated.

## Enzyme Activity of CmCCD4a

The enzymatic activities of CCDs are still a matter of debate. Floss et al. (*28*) reported different substrate specificities of CCD1 between *in vitro* and *in vivo* analyses. *In vivo* analysis using carotenoid-accumulating *Escherichia coli* cells showed that recombinant CmCCD4a specifically cleaves double bonds at the 9,10 (9′,10′) positions of β-carotene (Figure 4) (*24*); it does not cleave xanthophylls including zeaxanthin and lutein *in vitro*, but it does cleave the double bond at the 9,10 position of 8′-apo-β-caroten-8′-al. However, the *in planta* activity of CCD4 has not yet been elucidated.

We introduced *CmCCD4a* into yellow-flowered chrysanthemum cultivars and analyzed carotenoids in petals of transformants in order to determine enzyme activity of CmCCD4a *in planta*. Major carotenoids contained in the yellow petals of chrysanthemum are lutein and lutein epoxides (*29*). All of these carotenoids disappeared in petals of transformants and the petal color of transformants was white, suggesting that CmCCD4a cleave lutein and lutein epoxides *in planta* (Ohmiya et al. unpublished result). It is also possible that CmCCD4a has an activity to cleave intermediates of these carotenoids.

Carotenoid cleavage products in chrysanthemum petals have not yet been identified. Carotenoids may possibly be cleaved into small and colorless molecules. In contrast, some plants accumulate orange or red apocarotenoids and express unique petal colors. For example, various apocarotenoids such as β-ionone, 10′-apocaroten-10′-oic acid, and hydroxy 10′-apocaroten-10′-oic acid are found in brown petals of boronia (*Boronia megastigma*) (*30*, *31*). The orange petals of Crocosmia contain crocin (*32*) and those of California poppy (*Eschscholzia californica* Cham.) contain crocetin (*33*). The red stigma of crocus accumulate a unique mixture of apocarotenoids—crocetin glycosides, picrocrocin, and safranal—which are responsible for the color, taste, and aroma of the spice saffron (*34*).

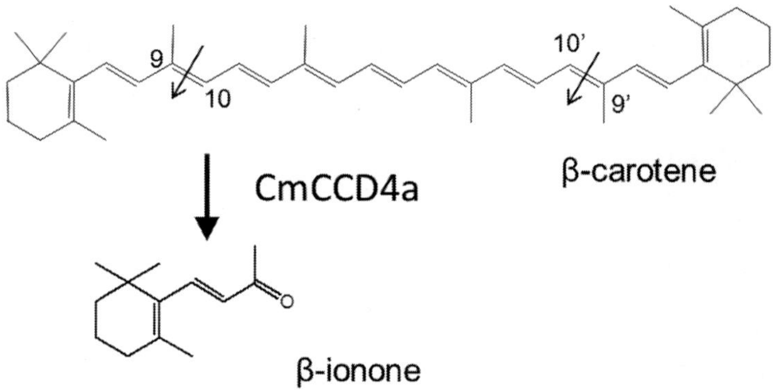


*Figure 4. Enzyme activity of CmCCD4a in β-carotene-producing E.coli, based on the results described by Huang et al. (24). However, enzyme activity of CmCCD4a in planta has not been fully elucidated.*

## White Petals Color Caused by Low Rate of Biosynthesis

To elucidate possible mechanisms that produce white petal color in other plants, expression of genes related to carotenoid metabolism were examined in white-flowered Japanese morning glory (*Ipomoea nil*) and carnation (*Dianthus caryophyllus* L.).

Although abundant flower colors of *I. nil* can be found, the flowers do not accumulate carotenoids and *I. nil* lacks a yellow-flowered cultivar. Despite a long history of attempts, crossbreeding aimed at producing yellow-flowered *I. nil* has never succeeded. Yamamizo et al. (*35*) showed that in petals of *I. nil,* expression levels of most of the carotenogenic genes functioning downstream of *isopentenyl pyrophosphate isomerase* (*IPI*) are extremely low; these genes include *geranylgeranyl pyrophosphate synthase*, *phytoene synthase* (*PSY*), *phytoene desaturase*, *ζ-carotene desaturase*, *carotenoid isomerase*, *lycopene ε-cyclase* (*LCYE*), *lycopene β-cyclase* (*LCYB*), and *β-ring hydroxylase*. The result suggests that, in petals of *I. nil*, flux through the carotenoid biosynthesis pathway is suppressed and carotenoid biosynthesis may not take place. Low rates of carotenoid biosynthesis were also postulated for carnation petals (Ohmiya et al. unpublished results). Microarray analysis showed that most of the carotenogenic genes were expressed in petals at levels similar to those in leaves and that expression levels increased during petal development. *PSY* and *LCYE* were exceptions. Their expression levels in petals were very low compared with those in leaves, and their expression decreased as petals matured, which is well correlated to the low carotenoid content. In many plants, PSY is a key enzyme that regulates the carotenoid level in the tissues (*36–38*). LCYE also plays an important role in the synthesis of α-carotene and lutein. Cunningham and Gantt (*39*) suggested that the apportioning of substrates into the pathways leading to α- and β-carotenes could be determined by the relative activities of LCYB and LCYE. Low expression levels of *PSY* and *LCYE* may cause a low rate of carotenoid biosynthesis in carnation petals.

To examine whether carotenoid degradation is responsible for the white color of *Ipomoea* petals, *CCD4* expression levels in petals were compared among white-flowered *I. nil* (no carotenoid), pale-yellow-flowered *Ipomoea obscura* (low carotenoid content), and yellow-flowered *Ipomoea* sp. (high carotenoid content). Substantial levels of *CCD4* expression were observed in petals of all of the *Ipomoea* plants examined and there was no correlation between the *CCD4* expression level and carotenoid content (*35*). White petal color would be dominant over the yellow color if carotenoid degradation is responsible for the white color formation of petals in *Ipomoea* plants. However, crossing experiments showed that the petal color of all the $F_1$ progenies obtained by crossing *Ipomoea* sp. with *I. obscura* are yellow (*35*), suggesting that carotenoid cleavage is not responsible for the low level of carotenoids in *I. obscura*. In the case of carnation, expression of *CCD4* was extremely low in petals (Ohmiya et al. unpublished results). These results indicate that, unlike the situation in chrysanthemum, the main reason why *I. nil* and carnation do not accumulate carotenoids in their petals is not rapid degradation but low rates of carotenoid biosynthesis.

## Future Perspective

The present paper shows that the carotenoid content in chrysanthemum petals is controlled by degradation, while low rate of biosynthesis is the cause of the absence of carotenoids in petals of *I. nil* and carnation. The results suggest that some plants use one of these strategies to control carotenoid accumulation in petals, and other plants use the other strategy.

There are many yellow flowers containing large amounts of carotenoids. On the other hand, there are many plant families or genera that do not accumulate carotenoids in their petals, such as Geraniaceae, Laminales, and Ericales. From evolutionary and physiological perspectives, it is of great interest to know why and how these plants do not accumulate carotenoids in their petals. However, our knowledge of flower carotenoids is limited and further work is required to answer these questions.

# References


1. Ohmiya, A. *Jpn. Agric. Res. Q. (JARQ)* **2011**, *45*, 163–171.
2. Machin, B.; Scope, N. In *Chrysanthemums: Year-Round Growing*; Machin, B., Scope, N., Eds.; Blandford Press: Dorset, U.K., 1978; pp 34–37.
3. Stewart, R. N.; Derman, H. *Amer. J. Bot.* **1970**, *57*, 1061–1071.
4. Jordan, C.; Reimann-Philipp, R. *Z. Pflanzenzuecht.* **1983**, *91*, 111–122.
5. Boase, M. R.; Miller, R.; Deroles, S. C. In *Plant Breeding Reviews*; Janic, J., Ed.; Wiley: New York, 1997; Vol. 14, pp 321–361.
6. Hattori, K. *Jpn. J. Breed.* **1991**, *41*, 1–9.
7. Ohmiya, A.; Kishimoto, S.; Aida, R.; Yoshioka, S.; Sumitomo, K. *Plant Physiol.* **2006**, *142*, 1193–1201.
8. Ohmiya, A.; Sumitomo, K.; Aida, R. *J. Jpn. Soc. Hortic. Sci.* **2009**, *78*, 450–455.
9. Ohmiya, A.; Toyoda, T.; Watanabe, H.; Emoto, K.; Hase, Y.; Yoshioka, Y. *J. Jpn. Soc. Hortic. Sci.* **2012**, *81*, 269–274.
10. Kishimoto, S.; Ohmiya, A. *Physiol. Plant.* **2006**, *128*, 437–447.
11. Yoshioka, S.; Sumitomo, K.; Fujita, Y.; Yamagata, A.; Onozaki, T.; Shibata, M.; Ohmiya, A. *Euphytica* **2010**, *171*, 295–300.
12. Yoshioka, S.; Aida, R.; Yamamizo, C.; Shibata, M.; Ohmiya, A. *Euphytica* **2012**, *184*, 377–387.
13. Simkin, A. J.; Schwartz, S. H.; Auldridge, M.; Taylor, M. G.; Klee, H. J. *Plant J.* **2004**, *40*, 882–892.
14. Simkin, A. J.; Underwood, B. A.; Auldridge, M.; Loucas, H. M.; Shibuya, K.; Schmelz, E.; Clark, D. G.; Klee, H. J. *Plant Physiol.* **2004**, *136*, 3504–3514.
15. Mathieu, S.; Terrier, N.; Procureur, J.; Bigey, F.; Günata, Z. *J. Exp. Bot.* **2005**, *56*, 2721–2731.
16. Baldermann, S.; Kato, M.; Kurosawa, M.; Kurobayashi, Y.; Fujita, A.; Fleischmann, P.; Watanabe, N. *J. Exp. Bot.* **2010**, *61*, 2967–2977.
17. Endo, N. *J. Jpn. Soc. Hortic. Sci.* **1969**, *38*, 343–349.
18. Shibata, M.; Kishimoto, S.; Hirai, M.; Aida, R.; Ikeda, I. *Acta Hortic.* **1998**, *454*, 347–353.
19. Langton, F. A. *Euphytica* **1980**, *29*, 807–812.
20. Shibata, M.; Kawata, J. In *Development of New Technology for Identification and Classification of Tree Crops and Ornamentals*; Kitaura, K., Akihama, T., Kukimura, H., Nakajima, K., Horie, M., Kozaki, I., Eds.; Fruit Tree Research Station, Ministry of Agriculture, Forestry and Fisheries: Japan, 1986; pp 41–45.
21. Dowrick, G. J.; El-Bayoumi, A. *Euphytica* **1966**, *15*, 32–38.
22. Tan, B. C.; Joseph, L. M.; Deng, W. T.; Liu, L.; Li, Q. B.; Cline, K.; McCarty, D. R. *Plant J.* **2003**, *35*, 44–56.
23. Rubio, A.; Rambla, J. L.; Santaella, M.; Gómez, M. D.; Orzaez, D.; Granell, A.; Gómez-Gómez, L. *J. Biol. Chem.* **2008**, *283*, 24816–24825.
24. Huang, F. C.; Molnár, P.; Schwab, W. *J. Exp. Bot.* **2009**, *60*, 3011–3022.
25. Campbell, R.; Ducreux, L. J. M.; Morris, W. L.; Morris, J. A.; Suttle, J. C.; Ramsay, G.; Bryan, G. J.; Hedley, P. E.; Taylor, M. A. *Plant Physiol.* **2010**, *154*, 656–664.

26. Brandi, F.; Bar, E.; Mourgues, F.; Horváth, G.; Turcsi, E.; Giuliano, G.; Liverani, A.; Tartarini, S.; Lewinsohn, E.; Rosati, C. *BMC Plant Biol.* **2011**, *11*, 24.
27. Eugster, C. H.; Márki-Fisher, E. *Angew. Chem. Int. Ed. Engl.* **1991**, *30*, 654–672.
28. Floss, D. S.; Schliemann, W.; Schmidt, J.; Strack, D.; Walter, M. H. *Plant Physiol.* **2008**, *148*, 1267–1282.
29. Kishimoto, S.; Maoka, T.; Nakayama, M.; Ohmiya, A. *Phytochemistry* **2004**, *65*, 2781–2787.
30. Cooper, C. M.; Davies, N. W.; Menary, R. C. *J. Agric. Food Chem.* **2003**, *51*, 2384–2389.
31. Cooper, C. M.; Davies, N. W.; Menary, R. C. *J. Agric. Food Chem.* **2009**, *57*, 1513–1520.
32. Ootani, S.; Hayashi, K. *Res. Inst. Evolut. Biol. Sci. Rep.* **1982**, *1*, 71–76 [In Japanese with English summary].
33. Wakelin, A. M.; Lister, C. E.; Conner, A. J. *Int. J. Plant Sci.* **2003**, *164*, 867–875.
34. Tarantilis, P. A.; Polissiou, M.; Manfait, M. *J. Chromatogr.* **1994**, *664*, 55–61.
35. Yamamizo, C.; Kishimoto, S.; Ohmiya, A. *J. Exp. Bot.* **2010**, *61*, 709–719.
36. von Lintig, J.; Welsch, R.; Bonk, M.; Giuliano, G.; Batschauer, A.; Kleinig, H. *Plant J.* **1997**, *12*, 625–634.
37. Giuliano, G.; Bartley, G. E.; Scolnik, P. A. *Plant Cell* **1993**, *5*, 379–387.
38. Welsch, R.; Beyer, P.; Hugueney, P.; Kleinig, H.; von Lintig, J. *Planta* **2000**, *211*, 846–854.
39. Cunningham, F. X., Jr; Gantt, E. *Proc. Natl. Acad. Sci. U.S.A.* **2001**, *98*, 2905–2910.

# Chapter 4

# Carotenoid Cleavage Dioxygenase and Presence of Apo-Carotenoids in Biological Matrices

**Rachel E. Kopec[1,2] and Steven J. Schwartz[*,1,2]**

**[1]Department of Food Science and Technology, The Ohio State University, Columbus, Ohio, United States**
**[2]Department of Human Nutrition, The Ohio State University, Columbus, Ohio, United States**
**[*]E-mail: schwartz.177@osu.edu. Address: 2015 Fyffe Road, 110 Parker Building, Columbus, OH 43210.**

Evidence exists demonstrating that carotenoid cleavage dioxygenase enzymes in plants exhibit broad substrate specificity for several carotenoid pigments. Non-specific phase I and phase II metabolism is also suggested to generate carotenoid metabolites *in vivo*. More recently in animals and humans, specific oxidation products of carotenoids have been identified. These products are hypothesized to possess biological activity similar to retinoids with potential to bind to the retinoid receptors in-vivo. Although oxidative eccentric cleavage mechanisms of carotenoids have been well documented, few reports have unequivocally identified the in-vivo apo-carotenoid and apo-lycopenoid metabolic products. Our laboratory has developed highly selective and sensitive liquid chromatography tandem mass spectrometry methodology to analyze these compounds in foods and biological tissues. In collaboration with other researchers at the Ohio State University, several eccentric cleavage products have been identified in plants and humans by comparison to synthesized authentic compounds, and their potential biological activity demonstrated.

## Introduction

In the past decade, the genes encoding nearly all the enzymes for carotenoid biosynthesis in plants have been identified, and their enzymatic activities have been characterized (*1–4*). In contrast, only recently have we begun to understand the function of carotenoid cleavage enzymes in plants. Likewise, metabolism of provitamin A carotenoids in humans, most notably β-carotene, has also been well studied because of the obvious nutritional importance. However, recent research has demonstrated that non-central cleavage products of lycopene and β-carotene are present in foods and in humans, and suggested that these products may play a biological role as well.

## Lycopene and Plant Carotenoid Cleavage Enzymes

To date, multiple carotenoid cleavage dioxygenase enzymes (CCD) have been identified *in Arabidopsis thaliana.* Of these, five enzymes are involved in the synthesis of abscisic acid, an important hormone regulating plant growth. The activity of the other CCD enzymes is still being characterized. CCD1 from maize and tomato has been shown to cleave lycopene at the 5′-6′ position and 9′-10′ position (*5*). While only the short, volatile products of these cleavages were determined by GC-MS (pseudo-ionone and 6-methyl-5-heptene-2-one), the complementary long-chain products of this enzymatic step would likely be apo-6′-lycopenal and apo-10′-lycopenal, respectively. In addition, carotenoid cleavage dioxygenase 1 (CCD1) in the annatto plant has been shown to cleave lycopene to produce bixin aldehyde, a precursor of the carotenoids bixin and norbixin (*6*). CCD7 appears to cleave the 9′-10′ double bond of lycopene, and the resulting product has been tentatively identified with UV spectra data and retention time (*7*). In addition to specific enzymatic cleavage by CCD enzymes, *in vitro* work has demonstrated that lycopene can be non-specifically co-oxidized with lipid by fruit lipoxygenase enzymes (*8*).

## Lycopene and Mammalian Carotenoid Cleavage Enzymes

The most widely characterized pathway of carotenoid metabolism is the formation of vitamin A by the cleavage of ß-carotene (or other pro-vitamin A carotenoids) at the central double bond by ß-carotene oxygenase 1 (BCO1) to form two molecules of retinal (*9*). However, research suggests that BCO1 does not metabolize lycopene to any significant extent (*10*). A second carotenoid cleaving enzyme, ß-carotene oxygenase 2 (BCO2), has also been identified (*11*). BCO2 has been shown to eccentrically cleave ß-carotene (*11*) and lycopene (*11*, *12*). Hu et al. (*12*) transfected COS-1 cells with ferret BCO2, and found that incubation with 5-(*Z*)-lycopene and 13-(*Z*)-lycopene (but not all-(*E*)-lycopene) produced apo-10′-lycopenal. Other researchers have incubated lycopene with various tissue homogenates *ex vivo* and produced oxidative products, but it is unclear whether these are products of a BCO reaction or whether they are produced *in vivo* (*8*).

## Lycopene and Phase I and Phase II Metabolism in Mammals

In a study by Zaripheh et al. (*13*), rats were fed lycopene as part of their daily diet for 50 days, followed by a radioactive dose of [$^{14}C$] labeled lycopene on day 51. Afterwards, radioactivity (as measured by scintillation counting) was observed in both the feces and the urine up to 168 hrs. post-dosing. While a majority of the excreted radioactivity was observed in feces, ~3% of the radioactivity observed was in urine (*13*). A similar study was recently conducted in humans, where [$^{14}C$] labeled lycopene was fed to two healthy subjects and samples of blood plasma, urine, skin, and breath $CO_2$ were collected (*14*). Accelerator mass spectrometry (AMS) was used to measure the radioactive dose. It was determined that 17%-19% of the radioactivity was excreted in the urine over 6 days after the lycopene dose consumption (*14*). Since lycopene is particularly hydrophobic, it is reasonable to hypothesize that it would need to be metabolized to more polar compounds in order to be excreted via urine. Metabolism by carotenoid cleavage enzymes provide one possible route to a more water-soluble metabolite. However, an alternative route creating such a product would be via metabolism by phase I and phase II enzymes. Ross et al. treated urine samples with β-glucuronidase and sulphatase to deconjugate potential phase II metabolites, but specific compounds were not identified, so it is unclear at this time if phase II metabolites were present (*14*).

Several investigations have been conducted to explore the hypothesis that dietary lycopene modulates the P450 enzymes. A study by Jewell & O'Brien (*15*) was conducted where male Wistar rats were fed a diet supplemented with lycopene (300 mg/kg diet) for 16 days. Afterwards, the activity of cytochrome P450 (CYP450) monoxygenase towards ethoxyresorufin (specific for P450 1A1), methoxyresorufin (specific for P450 1A2), pentoxyresorufin (specific for P450 2B1/2), benzyloxyresorufin (non-specific for P450 1A1/2, 2B1/2, and 3A) was determined in various tissues (liver, lung, kidney, small intestine). Gluthathione-S-transferase activity and glutathione levels were also measured. There was no difference observed in the activity of all enzymes measured or in glutathione levels between the lycopene supplemented group and the negative control group (*15*). A similar study performed by Gradelet et al. determining activity of the same four CYP450 enzymes also found no change in enzyme activity after lycopene supplementation (*16*). However, this group found that lycopene reduced the activity of nitrosodimethylamine N-demethylase (*16*). Similarly, a study by Liu et al. (*17*) found no change in CYP 1A1 activity or NAD(P)H:quinone reductase 1 activity (NQO1) in male Copenhagen rats fed 10% tomato powder for 7 days. In contrast, a study by Breinholt et al. (*18*) observed dose dependent increases in CYP activity towards ethoxyresorufin (specific for P450 1A1) and benzyloxyresorufin (non-specific for P450 1A1/2, 2B1/2, and 3A) in female Wistar rats fed a lycopene/corn oil solution via gavage daily for 14 days. A dose response effect of lycopene on activity of glutathione-S-transferase was observed, and interestingly, a U-shaped dose response curve was observed for other phase II enzymes investigated in this study (NQO1, UDP-glucuronosyltransferase) (*18*). O'Brien and O'Connor (*19*) have suggested that these different results may be due to differences in gender, species strain, or differences in levels of lycopene achieved *in vivo* (which was not measured in most studies cited above).

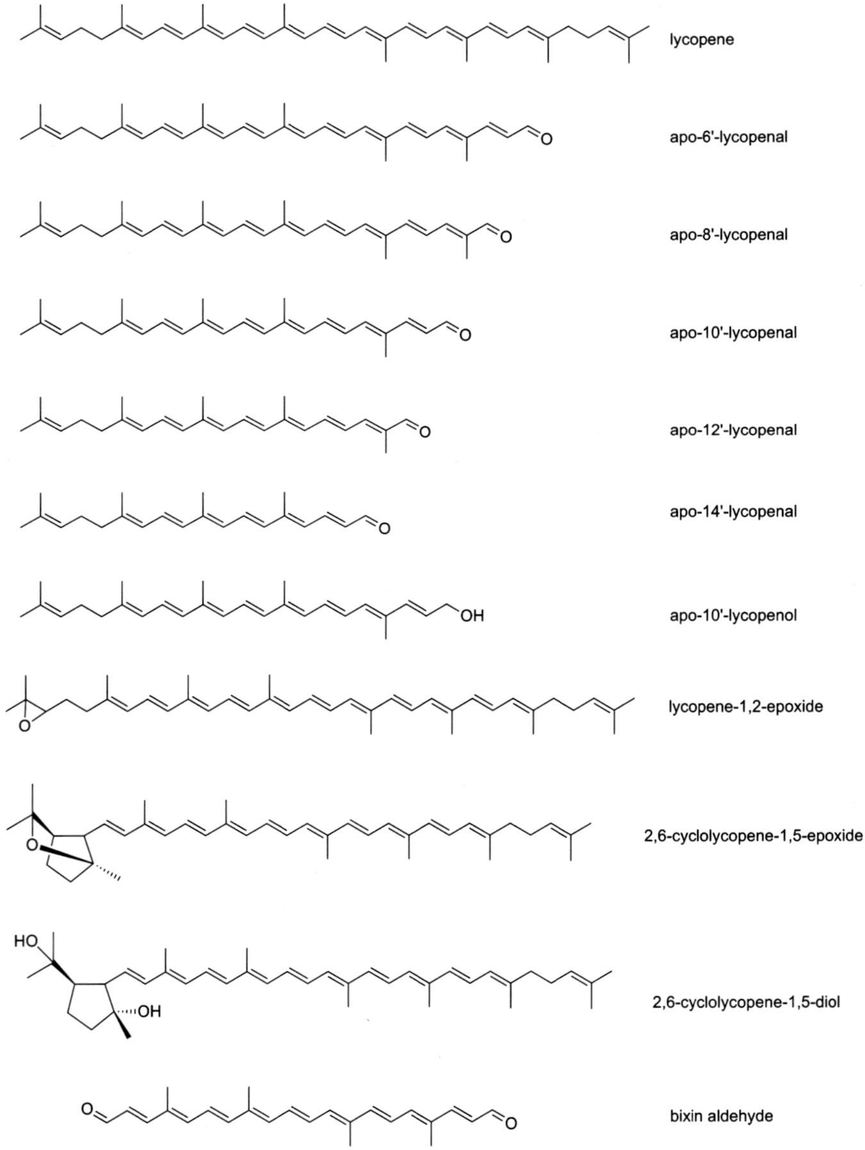


*Figure 1. Structures of lycopene and various lycopenoids, Taken from Story et al. (36). Reproduced with permission from reference (36). Copyright 2010 Annual Reviews.*

**Table 1. Apo-lycopenals in Commonly Consumed Lycopene Containing Foods.** ***Taken From Kopec et al., 2010 (25),*** **Reproduced with permission from reference (*25*). Copyright 2010 The American Chemical Society.**

| | *average µg/100g wet weight ± standard deviation* | | | | | |
|---|---|---|---|---|---|---|
| *Food* | *lycopene* | *apo-6'-lycopenal* | *apo-8'-lycopenal* | *apo-10'-lycopenal*[a] | *apo-12'-lycopenal* | *apo-14'-lycopenal*[b] |
| 'Roma' Tomato | 3,500 | 1.1 | 1.8 | 1.1 | 2.2 | 0.28 |
| Grape Tomato | 6,200 | 1.6 | 2.2 | 1.2 | 2.3 | 0.38 |
| Red Vine Ripened Tomato | 3,800 | 3.5 | 4.3 | 2.0 | 2.9 | 0.42 |
| Ruby Red Grapefruit | 190 | 0.023 | 0.027 | 0.022 | 0.047 | nd |
| Watermelon | 3,200 | 0.77 | 0.86 | 0.45 | 0.80 | 0.21 |
| Catsup[c] | 11,000 ± 1,600 | 4.9 ± 1.3 | 9.3 ± 1.6 | 1.2 ± 0.3 | 5.5 ± 0.2 | 0.27 ± 0.02 |
| Spaghetti Sauce[c] | 11,000 ± 2,700 | 6.3 ± 1.5 | 10 ± 2 | 1.2 ± 0.1 | 5.3 ± 0.8 | 0.31 ± 0.05 |
| Pizza Sauce[c] | 15,000 ± 4,100 | 10 ± 4 | 16 ± 7 | 1.7 ± 0.6 | 7.1 ± 2.2 | 0.43 ± 0.11 |
| Tomato Soup[c] | 9,600 ± 2,400 | 6.9 ± 3.7 | 11 ± 4 | 1.7 ± 0.8 | 6.1 ± 2.3 | 0.54 ± 0.27 |
| Tomato Paste[c] | 34,000 ± 4,100 | 19 ± 5 | 34 ± 3 | 3.7 ± 1.2 | 16 ± 4 | 0.69 ± 0.13 |
| Tomato Juice[c] | 8,400 ± 1,800 | 1.8 ± 0.5 | 5.0 ± 1.3 | 0.56 ± 0.13 | 2.8 ± 0.3 | 0.11 ± 0.02 |

nd = not detected [a] Level was estimated by averaging apo-8′-lycopenal and apo-12′-lycopenal slopes. [b] Level was estimated by using apo-12′-lycopenal equivalents. [c] Standard deviation reflects variation between three different commercial manufacturers of the same food product.

One human study observed the difference in CYP1A2 activity between individuals with colorectal cancer (n=43) and healthy controls (n=47) in relation to plasma carotenoid levels (*20*). In this study, plasma lycopene was positively associated with CYP1A2 activity ($p = 0.06$). Overall, evidence suggests that lycopene may be metabolized in mammals, either by enzymatic cleavage from BCO2 or from metabolism via one or more CYP enzymes. Furthermore, evidence suggests that lycopene metabolites may be conjugated to more polar products via phase II enzymes. However, much work still remains to identify the specific mechanisms and enzymes involved.

## Lycopene Isomerization After Feeding

Lycopene is primarily present in the all-(*E)* form in most foods, including tomatoes and tomato products (*21*). However, once absorbed into the human body, all-(*E*)-lycopene is converted into (*Z*)-lycopene, and steady state circulating blood levels show higher levels of (*Z*)-lycopene as compared to (*E)*-lycopene after a washout-phase (*22*). Recently, work in our laboratory, in collaboration with others, with stable isotopes suggests that isomerization takes place after absorption. Results of these studies will be published soon. Work by Hu et al. (*12*) has demonstrated that (*Z*)-lycopene is preferentially cleaved by BCO2 *in vitro*.

## Lycopene Oxidation Products in Plants and Animals

Several lycopenoids (a term used to describe oxidative metabolites of lycopene) have been identified in raw tomatoes and tomato products (Figure 1).

Apo-6′-lycopenal and apo-8′-lycopenal, were reported in tomato paste (*23*). Later in 1973, Ben Aziz et al. (*24*) confirmed the presence of apo-6′-lycopenal and apo-8′-lycopenal in raw tomatoes, indicating that these products are present in the unprocessed fruit as well. Further work by our group has demonstrated that in fact a whole series of apo-lycopenals are present in lycopene containing fruits and vegetables, and processed food products from these raw fruits and vegetables, as shown in Table 1, Kopec et al. (*25*).

Khachik et. al. first reported lycopene derivatives in human blood plasma (*26*), which were later found to be present in breastmilk, and identified as two diasteromeric forms of 2,6-cyclolycopene-1,5-diol (*27*). More recently, lycopenoids have been identified in the tissues of animals consuming lycopene. In one study, ferrets consumed a lycopene supplement (lycovit powder, 10% lycopene) daily for 9 weeks (*12*). Following sacrifice, apo-10′-lycopenol, and alcohol, was found in the lung tissue of these animals (*12*). These researchers went on to incubate ferret liver homogenate ex vivo with apo-10′-lycopenal (an aldehyde), and discovered that apo-10′-lycopenoic acid was produced when $NAD^+$ was added as a cofactor (*12*). In a separate study by Zaripheh et al. (*13*), F344 rats were fed lycopene as part of their daily diet for 50 days, followed by a radioactive dose of lycopene on day 51. Various tissues were extracted and analyzed by high performance liquid chromatography (HPLC)-photo diode array (PDA). HPLC fractions were collected and further analyzed by scintillation counting (*13*). Polar products, defined as products containing radioactivity which eluted in the first 5 min. of the HPLC run, were found in multiple tissues. Polar products were observed in liver (20% of isotope label), seminal vesicle (38% of isotope label), and prostate (67% of isotope label). In contrast, it is noted that polar products in the kidney, adrenal glands, and feces did not change over time, but it is unclear what % of the total isotope label was attributed to polar products in these tissues (*13*). Upon further analysis, the rat livers from the Zaripheh et al. (*13*) study were pooled and extracted with non-polar solvent, and apo-8′-lycopenal and putative apo-12′-lycopenal were identified (*28*). Apo-lycopenals have also been found in humans. In a study by our group, subjects consumed 300 mL of a tomato juice beverage daily for two months. Blood plasma samples were taken, extracted with non-polar solvents, and analyzed using high performance liquid chromatography-tandem mass spectrometry (HPLC-MS/MS), as shown in Figure 2.

Apo-lycopenals in the blood of seven individuals were present at the nanomolar level, as compared to lycopene which was present at the micromolar level (*25*). Research has suggested that some of these products may have biological activity. Apo-lycopenals have been show to activate the electrophile/antioxidant response element transcription system (*29*), and to reduce breast and prostate cancer cell growth *in vitro* (*29*).

Further work is needed to determine if these apo-lycopenal products are formed *in vivo*, whether they are absorbed from lycopene-containing foods, and if/how they may exert biological activity.

## Noncentral Metabolites of β-Carotene

Interest in non-provitamin A metabolites of β-carotene has also increased in recent years. Non-central metabolites of β-carotene have been reported in raw fruits such as melons (*30*), mamey fruit (*31*), and passionfruit (*32*). Likewise, non-central cleavage products of β-carotene have been identified in mice (*33*) and ferrets (*12*) as well.

Certain β-carotene oxidation products have also been identified in humans. β -apo-14′-carotenal was identified in the blood plasma of hemodialysis patients (*34*). This analyte was shown to reduce peroxisome proliferator activated receptor (PPAR) and retinoid X receptor (RXR) activation in *in vitro* (*34*). Further work by Eroglu et al., demonstrated that β -apo-13-carotenone, β -apo-14′-carotenal, and β -apo-14′-carotenoic acid acted as nuclear receptor agonists for the retinoic acid receptors (RAR) α, β, and γ *in vitro* (*35*). Modeling work suggests that these analytes compete with the natural agonist retinoic acid for binding in the active site of RAR (Figure 3). Furthermore, β -apo-13-carotenone was found in the blood plasma of six free living humans at ~4 nmolar range (*35*). This level is comparable to the level of circulating retinoic acid. Interestingly, β -apo-14′-carotenal and β -apo-14″-carotenoic acid were not found in these human plasma samples (*35*). Taken together, this work suggests that non-central cleavage products of β-carotene may be absorbed from foods and/or formed by carotenoid cleavage enzymes *in vivo*, and may exert biological activity by binding to nuclear receptors.

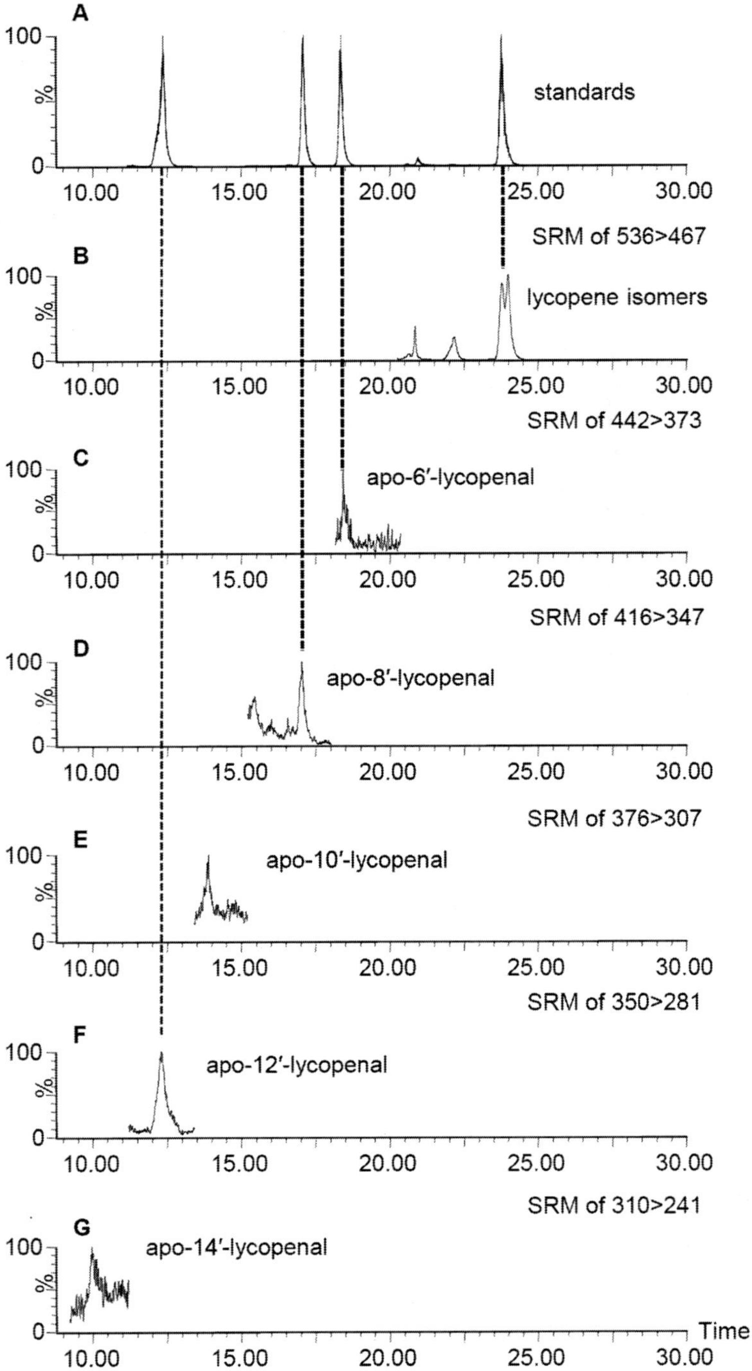

*Figure 2. HPLC-MS/MS analysis of lycopene and apo-lycopenals in the blood plasma extract of one representative individual. (A) MS chromatogram of the collection of SRM channels for all-(E)-lycopene (536 > 467), apo-6'- (442 > 373), apo-8'- (416 > 347), and apo-12'-lycopenal (350 > 281) standards and (B-G) lycopene and apo-lycopenals in blood plasma. Taken from Kopec et al., (25). Reproduced with permission from reference (25). Copyright 2010 The American Chemical Society.*

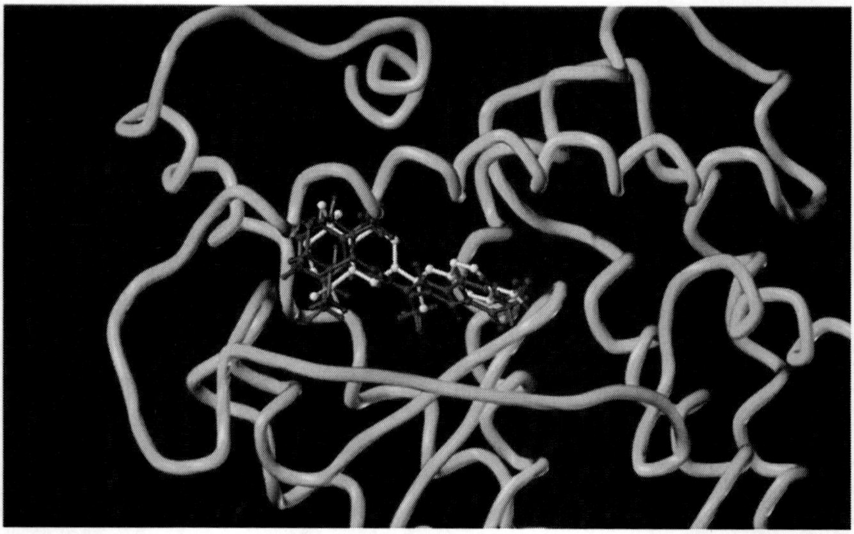

*Figure 3. Molecular model showing the binding of all-(E)- retinoic acid (red), β-apo-13-carotenone (purple) and the agonist tetramethyl tetrahydronaphthalenyl propenyl benzoic acid (TTNPB) (white) into the ligand binding site of the retinoic acid receptor (protein backbone in green). Taken from Eroglu et al., (35). Reproduced with permission from reference (35). Copyright 2012 The American Society for Biochemistry and Molecular Biology. (see color insert)*

## Conclusion

Research into the metabolism of carotenoids has recently revealed new enzymes and metabolite products both in plants and in mammals. Lycopene and β-carotene metabolites have been identified in plants, animals, and humans. These products may arise from cleavage by specific carotenoid cleavage enzymes, or other non-specific enzymatic or chemical processes. Regardless of how these products arise, their presence in humans indicates that they have the potential to mediate biological reactions *in vivo*. For example, recent evidence has suggested that lycopene and non-provitamin A metabolites of β-carotene may bind to nuclear receptors, and act as agonists or competitive antagonists to affect a host of downstream effects at the cellular level. Ultimately, these products may be responsible for reduced chronic disease risk associated with the consumption of carotenoid-containing fruits and vegetables. However, much work remains to be done before any firm conclusions can be reached.

## References


1. Cunningham, F. X.; Gantt, E. *Annu. Rev. Plant Physiol. Plant Mol. Biol.* **1998**, *49*, 557–583.
2. Hirschberg, J. *Curr. Opin. Plant Biol.* **2001**, *4*, 210–218.
3. Howitt, C. A.; Pogson, B. J. *Plant Cell Environ.* **2006**, *29*, 435–445.
4. Tanaka, Y.; Sasaki, N.; Ohmiya, A. *Plant J.* **2008**, *54*, 733–749.
5. Vogel, J. T.; Tan, B. C.; Mccarty, D. R.; Klee, H. J. *J. Biol. Chem.* **2008**, *283*, 11364–11373.
6. Bouvier, F.; Dogbo, O.; Camara, B. *Science (Washington DC)* **2003**, *300*, 2089–2091.
7. Schwartz, S. H.; Qin, X.; Loewen, M. C. *J. Biol. Chem.* **2004**, *279*, 46940–46945.
8. dos Anjos Ferreira, A. L.; Yeum, K.-J.; Russell, R. M.; Krinsky, N. I.; Tang, G. *J. Nutr. Biochem.* **2004**, *15*, 493–502.
9. von Lintig, J.; Vogt, K. *J. Biol. Chem.* **2000**, *275*, 11915–11920.
10. Lindqvist, A.; Andersson, S. *J. Biol. Chem.* **2002**, *277*, 23942–23948.
11. Kiefer, C.; Hessel, S.; Lampert, J. M.; Vogt, K.; Lederer, M. O.; Breithaupt, D. E.; von Lintig, J. *J. Biol. Chem.* **2001**, *276*, 14110–14116.
12. Hu, K. Q.; Liu, C.; Ernst, H.; Krinsky, N. I.; Russell, R. M.; Wang, X. D. *J. Biol. Chem.* **2006**, *281*, 19327–19338.
13. Zaripheh, S.; Boileau, T. W. M.; Lila, M. A.; Erdman, J. W. J. *J. Nutr.* **2003**, *133*, 4189–4195.
14. Ross, A. B.; Vuong, L. T.; Ruckle, J.; Synal, H. A.; Schulze-König, T.; Wertz, K.; Rümbeli, R.; Liberman, R. G.; Skipper, P. L.; Tannenbaum, S. R.; Bourgeois, A.; Guy, P. A.; Enslen, M.; Nielsen, I. L. F.; Kochhar, S.; Richelle, M.; Fay, L. B.; Williamson, G. *Am. J. Clin. Nutr.* **2011**, *93*, 1263–1273.
15. Jewell, C.; O'Brien, N. M. *Br. J. Nutr.* **1999**, *81*, 235–242.
16. Gradelet, S.; Astorg, P.; Leclerc, J.; Chevalier, J.; Vernevaut, M. F.; Siess, M. H. *Xenobiotica* **1996**, *26*, 49–63.
17. Liu, A. G.; Volker, S. E.; Jeffery, E. H.; Erdman, J. W. *J. Agric. Food Chem.* **2009**, *57*, 7304–7310.
18. Breinholt, V.; Lauridsen, S. T.; Daneshvar, B.; Jakobsen, J. *Cancer Lett.* **2000**, *154*, 201–210.
19. O'Brien, N.; O'Connor, T. In *Carotenoids in Health and Disease*; Krinsky, N. I., Mayne, S. T., Sies, H., Eds.; Marcel Dekker: New York, 2004; pp 197–208.

20. Le Marchand, L.; Franke, A. A.; Custer, L.; Wilkens, L. R.; Cooney, R. V. *Pharmacogenetics* **1997**, *7*, 11–19.
21. Nguyen, M. L.; Schwartz, S. J. In *Natural Food Colorants: Science and Technology*; Lauro, G., Francis, F. J., Eds.; Marcel Dekker: New York, 2000; pp 153−192.
22. Fröhlich, K.; Kaufmann, K.; Bitsch, R.; Böhm, V. *Br. J. Nutr.* **2006**, *95*, 734–741.
23. Winterstein, A.; Studer, A.; Rüegg, R. *Chem. Ber.* **1960**, *93*, 2951–2165.
24. Ben-Aziz, A.; Britton, G.; Goodwin, T. W. *Phytochemistry* **1973**, *12*, 2759–2764.
25. Kopec, R. E.; Riedl, K. M.; Harrison, E. H.; Curley, R. W., Jr; Hruszkewycz, D. P.; Clinton, S. K.; Schwartz, S. J. *J. Agric. Food Chem.* **2010**, *58*, 3290–3296.
26. Khachik, F.; Beecher, G. R.; Goli, M. B.; Lusby, W. R.; Smith, J. C. J. *Anal. Chem.* **1992**, *64*, 2111–2122.
27. Khachik, F.; Steck, A.; Niggli, U. A.; Pfander, H. *J. Agric. Food Chem.* **1998**, *46*, 4874–4884.
28. Gajic, M.; Zaripheh, S.; Sun, F.; Erdman, J. W. *J. Nutr.* **2006**, *136*, 1552–1557.
29. Linnewiel, K.; Ernst, H.; Caris-Veyrat, C.; Ben-Dor, A.; Kampf, A.; Salman, H.; Danilenko, M.; Levy, J.; Sharoni, Y. *Free Radic. Biol. Med.* **2009**, *47*, 659–667.
30. Fleshman, M. K.; Lester, G. E.; Riedl, K. M.; Kopec, R. E.; Narayanasamy, S.; Curley, R. W.; Schwartz, S. J.; Harrison, E. H. *J. Agric. Food Chem.* **2011**, *59*, 4448–4454.
31. Godoy, H. T.; Rodriguez-Amaya, D. B. *J. Agric. Food Chem.* **1994**, *42*, 1306–1313.
32. Luenberger, F. J. *Z. Lebensm.-Unters. Forsch.* **1972**, *149*, 279–282.
33. Shmarakov, I.; Fleshman, M. K.; D'Ambrosio, D. N.; Piantedosi, R.; Riedl, K. M.; Schwartz, S. J.; Curley, R. W., Jr; von Lintig, J.; Rubin, L. P.; Harrison, E. H.; Blaner, W. S. *Arch. Biochem. Biophys.* **2010**, *504*, 3–10.
34. Ziouzenkova, O.; Orasanu, G.; Sukhova, G.; Lau, E.; Berger, J. P.; Tang, G.; Krinsky, N. I.; Dolnikowski, G. G.; Plutzky, J. *Mol. Endocrinol.* **2007**, *21*, 77–88.
35. Eroglu, A.; Hruszkewycz, D. P.; dela Sena, C.; Narayanasamy, S.; Riedl, K. M.; Kopec, R. E.; Schwartz, S. J.; Curley, R. W., Jr; Harrison, E. H. *J. Biol. Chem.* **2012**, *287*, 15886–15895.
36. Story, E.; Kopec, R. E.; Schwartz, S. J.; Harris, G. K. *Nutr. Rev.* **2010**, *1*, 189–210.

# Aroma-Active Carotenoid Metabolites

## Chapter 5

# Carotenoid Cleavage Products in Saffron (*Crocus sativus* L.)

**Ana M. Sánchez and Peter Winterhalter***

**Institute of Food Chemistry, Technische Universität Braunschweig, Schleinitzstrasse 20, 38106 Braunschweig, Germany**
***E-mail: p.winterhalter@tu-bs.de.**

Saffron (*Crocus sativus* L.) is cultivated for the color, taste, aroma, and biological properties of its stigmas, which once dehydrated give rise to the most expensive spice in the world. Responsible for these valued properties are several carotenoid cleavage products, such as crocins, picrocrocin, and safranal. Crocins, the major pigments of saffron spice, are a group of water-soluble glycosyl esters of crocetin (8,8′-diapo-ψ,ψ′-carotenedioic acid). Picrocrocin (4-(β-D-glucopyranosyloxy)-2,6,6-trimethyl-1-cyclohexene-1-carboxaldehyde) is the fore-most contributor to its pleasant bitter taste; and safranal (2,6,6-trimethyl-1,3-cyclohexadiene-1-carboxaldehyde) stands out in its volatile fraction. In this work, a review concerning these major and several minor carotenoid cleavage products in saffron is presented. Aspects like spectral features, precursors, biosynthesis, glucosylation and cyclization as well as the enzymes involved are discussed together with the functions and importance of these carotenoid-derived products.

# Introduction

Saffron is the species *Crocus sativus* L. and is phylogenetically classified in the clade Angiosperms, clade *Monocots*, order *Asparagales*, family *Iridaceae*, subfamily *Crocoideae* and genus *Crocus* (*1*, *2*). It is taxonomically classified in the division *Spermatophyta,* subdivision *Magnoliophytina* (Angiosperms), class *Liliatae* (Monocotyledons), subclass *Liliidae,* order *Liliales,* family *Iridaceae* and genus *Crocus* (*3*). Saffron is a triploid perennial sterile plant, which has been reproduced vegetatively for millennia by corms (*4*). Saffron is cultivated to produce saffron spice, the most expensive spice in the world. This spice consists of the dried stigmas of saffron blossoms and stands out of its color, taste, aroma and biological properties.

According to statistics from the Iranian ministry of agriculture (*5*), world production of saffron spice was about 245 tons in 2005 and Iran was the main producer (93.7% of the world wide production) and exporter (188.6 tons, mostly in bulk). India, Greece, Spain, Morocco and Italy are also involved in saffron production and play important roles in the trade of saffron spice. In particular Spain, which was an important producer in the past, leads saffron spice trade through processing and re-exporting saffron produced in other countries with an extraordinary final product quality.

Three major groups of carotenoid derivatives –a) glycosyl esters of crocetin (8,8′-diapo-ψ,ψ′-carotenedioic acid) (Figure 1), b) non volatile glycosidic aroma precursors like picrocrocin (4-(β-D-glucopyranosyloxy)-2,6,6-trimethyl-1-cyclohexene-1-carboxaldehyde) (Figure 2) and c) volatile compounds, mainly safranal (2,6,6-trimethyl- 1,3-cyclohexadiene-1-carboxaldehyde) (Figure 3)– are to a great extent responsible for the sensorial properties of saffron. The levels of these carotenoid derivatives are of critical importance because they determine the quality and commercial value of the plant. It is known that their levels depend on climatic conditions, harvesting, stigma separation, handling, storage, packaging and, above all, the drying process (*6–8*).

Two main mechanisms of carotenoid catabolism can be distinguished: a non-specific one by (photo)chemical oxidation or oxidation by non-specific enzymes like lipoxygenases and peroxidases; or a specific cleavage by enzymatic action of a family of oxidative enzymes. The first protein found to specifically cleave carotenoids (VP14) was identified in maize (*9*, *10*). This work contributed to the discovery of related enzymes in other plants and organisms. In *Arabidopsis thaliana* nine clades of dioxygenases were defined (*11*). Five of these clades, the 9-*cis* epoxycarotenoid dioxygenases (NCEDs): NCED2, NCED3, NCED5, NCED6, and NCED9, are closely related to VP14 and also cleave the 11,12(11′,12′) double bonds of the 9-*cis* isomers of neoxanthin and violaxanthin to yield xanthoxin, the precursor of abscisic acid (ABA). The remaining four clades: CCD1, CCD4, CCD7, and CCD8) have low sequence homologies, and their enzyme activity and substrate specificity differ when comparing to the NCEDs. Orthologs belonging to the NCED and CCD subfamilies have been identified in many other plant species, and they are named on the basis of their homology to those of *Arabidopsis* (*12*).

Advances in the chemical characterization of saffron with the detection of novel carotenoid metabolites and advances in its molecular and genetic bases are summarized in this chapter.

## Structure, Function, and Levels of Carotenoid Cleavage Products

### Crocetin Esters

The structures of the crocetin and the crocetin esters reported in saffron until now are shown in Figure 1. These highly water-soluble compounds are responsible for the red color of saffron stigmas and saffron spice and the yellow hues of its solutions. In fact, color is the most important quality parameter of saffron spice and the coloring strength values of its aqueous extracts are critical for the commercial value of the spice. *Cis*-crocetin esters have, compared to their *trans*-configured isomers, a lower fusion point, an additional absorption band at about 324 nm and a maximum at about 435 nm instead of at 440 nm in their UV-vis spectrum (*13–16*). During its development and growth, the stigma changes in color from white to scarlet, passing through yellow, orange and red stages. These stages were defined as follows in order to study the accumulation of color and aroma compounds during stigma development (*17*, *18*): yellow stigma, closed bud inside the perianth tube (around 0.3 cm in length); orange stigma, closed bud inside the perianth tube (around 0.4 cm in length); red stigma, closed bud inside the perianth tube (around 0.8 cm in length); -3da, 3days before anthesis, dark red stigmas in closed bud outside the perianth tube (3 cm); da, day of anthesis, dark red stigmas (3 cm); +2da, 2 days after anthesis, dark red stigmas; scarlet stage, from -3da to +2da. Castillo et al. (*17*) investigated the accumulation and the molecular mechanisms that regulate the synthesis of crocetin and crocin during stigma development. The red stage presented the highest level of crocetin and crocetin esters. The mono- and diglucosidic crocetin esters also reached a maximum in the red stage, whereas higher glucosylated crocetin esters reached the highest levels in the scarlet stages. Moreover, accumulation of crocetin esters of higher glucose content was in line with the expression patterns of UGTCs2, a glucosyltransferase involved in crocetin and crocetin esters glucosylation in stigmas (*19*).

Aspects such as crocetin esters chemical characterization, analysis, degradation and biosynthesis were reviewed by Carmona et al. (*20*). In addition, the authors reported the influence of the dehydration process on crocetin ester content, the relationship between visual color and coloring strength, the evolution of crocetin esters during accelerated aging, and their contribution to the aroma generation. The profile of crocetin esters is characteristic of saffron spice and is influenced by the different production processes. Crocetin ester contents ranging from 0.5 to 32.4 % on a dry basis have been reported (*21*) for saffron spice.

Gardenia fruit (*Gardenia jasminoides* Ellis) is another source of crocetin esters and thus, a potential substitute or adulterant of saffron. A natural yellow colorant also known as gardenia is obtained from this fruit and used in food and oriental medicine (*22*).

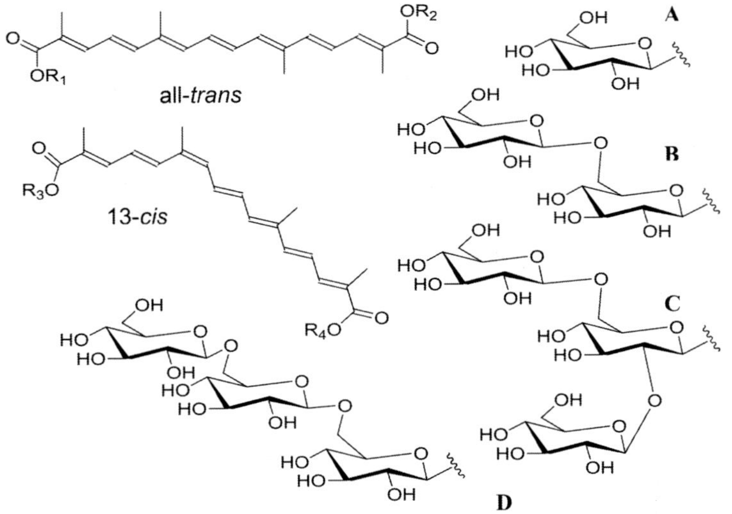


| | $R_1$ | $R_2$ |
|---|---|---|
| Crocetin | H | H |
| *Trans*-crocetin (β-D-glucosyl) ester | H | A |
| *Trans*-crocetin di-(β-D-glucosyl) ester | A | A |
| *Trans*-crocetin (β-D-gentiobiosyl) ester | H | B |
| *Trans*-crocetin (β-D-glucosyl)-( β-D-gentiobiosyl) ester | A | B |
| *Trans*-crocetin di-(β-D-gentiobiosyl) ester | B | B |
| *Trans*-crocetin (β-D-glucosyl)-( β-D-neapolitanosyl) ester | A | C |
| *Trans*-crocetin (β-D-gentiobiosyl)-( β-D-neapolitanosyl) ester | B | C |
| *Trans*-crocetin (β-D-triglucosyl)-( β-D-neapolitanosyl) ester | D | C |

| | $R_3$ | $R_4$ |
|---|---|---|
| *Cis*-crocetin di-(β-D-glucosyl) ester | A | A |
| *Cis*-crocetin (β-D-gentiobiosyl) ester | H* | B* |
| *Cis*-crocetin (β-D-glucosyl)-( β-D-gentiobiosyl) ester | A* | B* |
| *Cis*-crocetin di-(β-D-gentiobiosyl) ester | B | B |
| *Cis*-crocetin (β-D-glucosyl)-( β-D-neapolitanosyl) ester | A* | C* |
| *Cis*-crocetin (β-D-gentiobiosyl)-( β-D-neapolitanosyl) ester | B* | C* |
| *Cis*-crocetin (β-D-triglucosyl)-( β-D-neapolitanosyl) ester | D* | C* |

*Figure 1. Structures of crocetin and crocetin esters (*exchangeable).*

## Picrocrocin and Other Nonvolatile Glycosidic Aroma Precursors

Picrocrocin (Figure 2) is considered to be the main contributor to the bitter taste of saffron. It is also the precursor of safranal, the major compound in the saffron spice volatile fraction contributing to its aroma. During stigma development the levels of picrocrocin reached a maximum at anthesis, coinciding with the moment of flower collection for saffron spice production (*18*). In addition to picrocrocin, other non-volatile aroma precursor forms are present in the spice (*23–25*) (cf. Figure 2). The reported contents of picrocrocin in saffron spice ranged from 0.8 to 26.6 % on a dry basis (*21*, *26*, *27*).

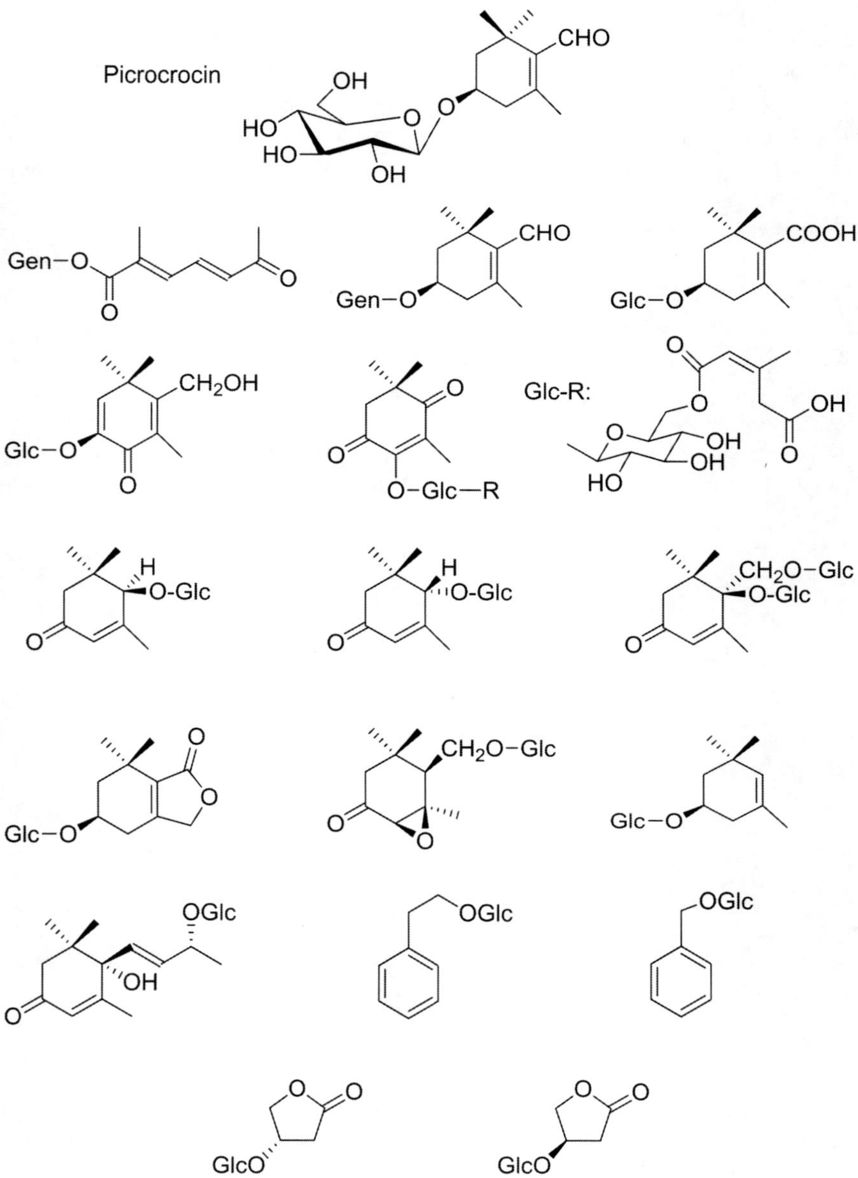


*Figure 2. Picrocrocin and other non-volatile glycoconjugates of saffron.*

## Safranal and the Aroma of Saffron

Kuhn and Winterstein obtained safranal, the characteristic impact compound of saffron spice aroma, as a degradation product of picrocrocin already in 1934 (*28*). Since then, many other constituents have been identified in the volatile fraction of the spice that contribute to its typical aroma even when only traces of them are present (Figure 3) (*29*). These compounds contribute odor notes such as e.g. caramel-like, citrus, spicy, flower, corn, roasty, sweet, earthy, nutty, popcorn, and baked potato (*30*, *31*). Many of these volatiles are not present in fresh saffron stigmas and are generated during the procurement process of saffron spice or even after cooking from degradation of volatile and non-volatile precursor compounds.

Two recent works of the group of Gómez-Gómez (*18*, *32*) has dealt with the volatiles emitted by fresh saffron stigmas during their development. Their results showed an early but low emission at the first stages of development and an increment of volatile production until the anthesis (complete opening of the flower). There was a notable change in volatile composition as stigmas developed with an important production of carotenoid-derivative volatiles together with fatty acid derivatives and terpenoids. With regard to carotenoid-derivative volatiles, in the yellow, orange and red stages, apocarotenoids $C_{13}$ volatiles (6,10-dimethyl-3,5,9-undecatrien-2-one (pseudoionone), α-ionone, 4-(2,6,6,trimethyl-cyclohexa-1,3-dienyl)but-2-en-4-one (damascenone)) were nearly undetectable. 6,10-Dimethyl-5,9-undecadien-2-one (geranylacetone) and 6-methyl-5-hepten-2-one (MHO) showed low levels in all stages, whereas β-cyclocitral, the 7,8 cleavage product of β-carotene, was highly produced in red stigmas.

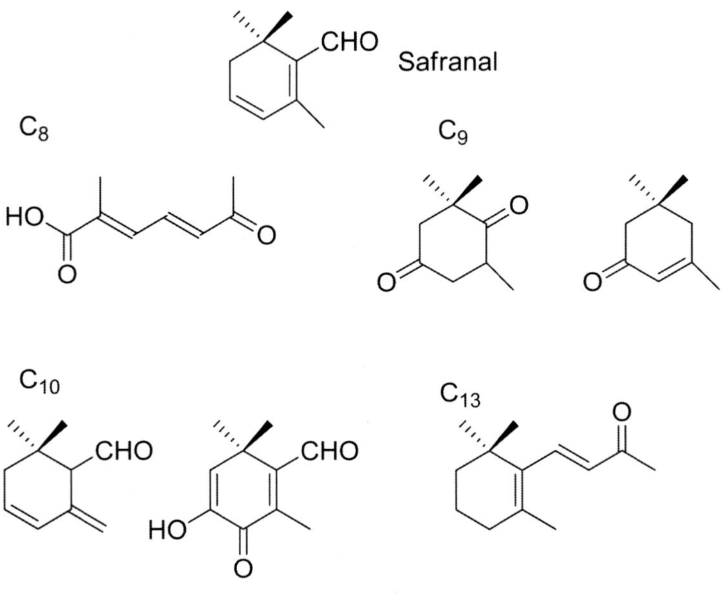


*Figure 3. Safranal and examples of other volatile $C_8$-, $C_9$-, $C_{10}$-, and $C_{13}$ -carotenoid breakdown products in saffron.*

**Table I. Enzymes Involved in the Biosynthesis of Saffron Carotenoids Including the Substrates, Activities, and Products**

| *Enzyme* | *Substrate and activity* | *Product* |
| --- | --- | --- |
| *CsPSY* | Geranylgeranylpyrophosphate<br>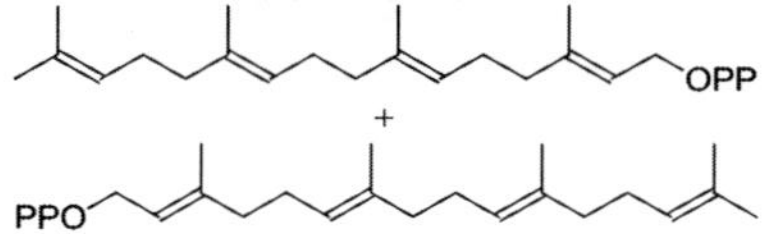<br><br>Coupling | 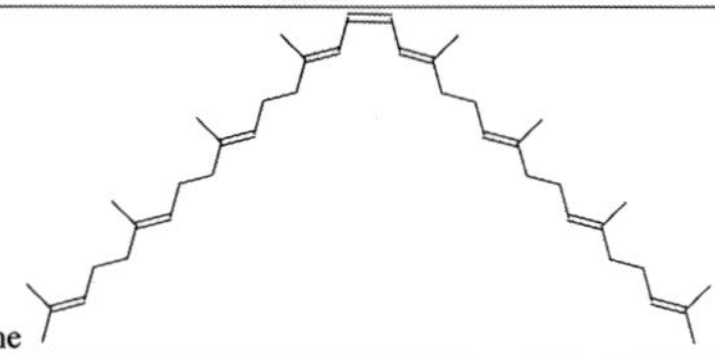<br>15-*cis*-Phytoene |
| *CsPDS* | 15-*cis*-Phytoene<br>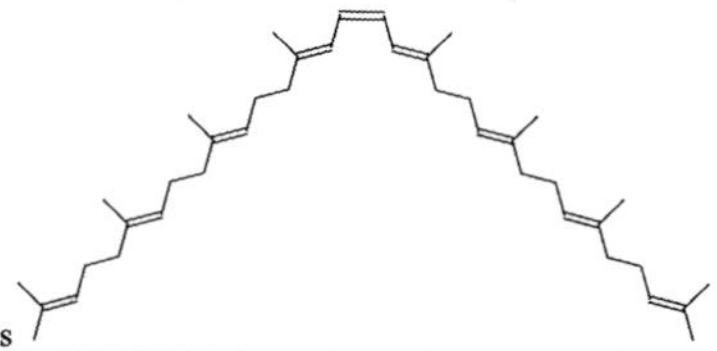<br>Two sequential desaturations | 9,15,9'-tri-*cis*-ζ-Carotene<br>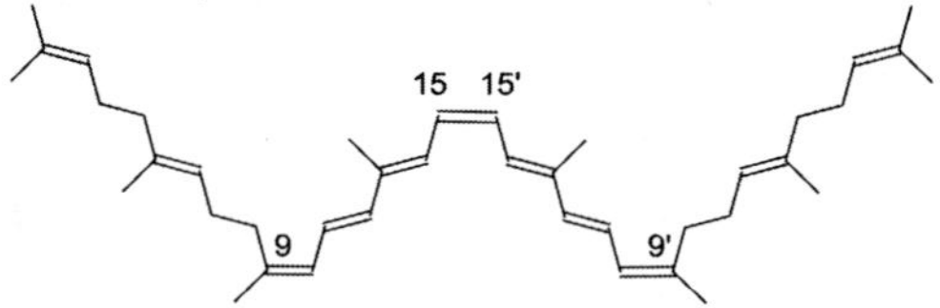<br> |
| *CsβLCY* | all-*trans* Lycopene<br>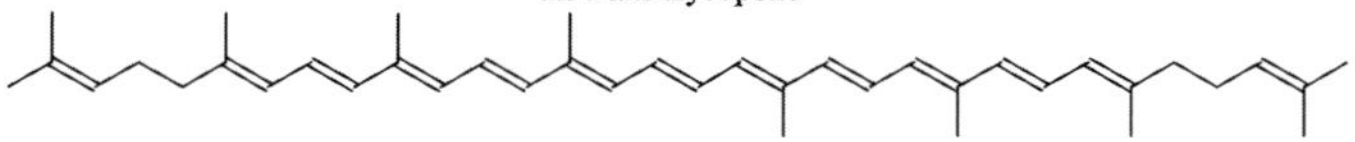<br>Cyclization | β-Carotene<br>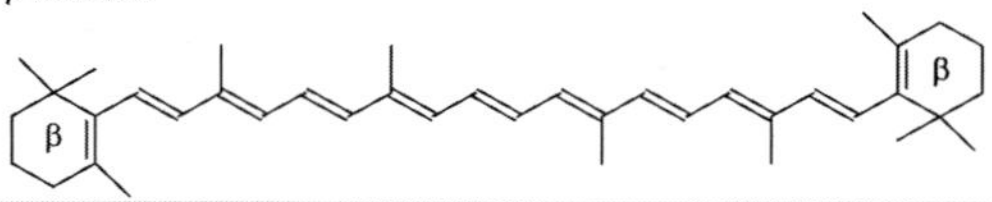<br> |
| *CsBCH* | β-Carotene<br>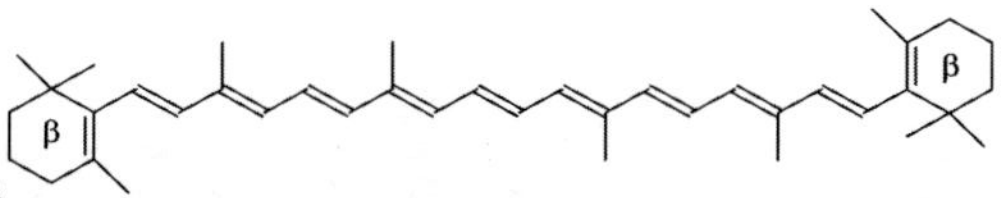<br><br>Hydroxylation | Zeaxanthin<br>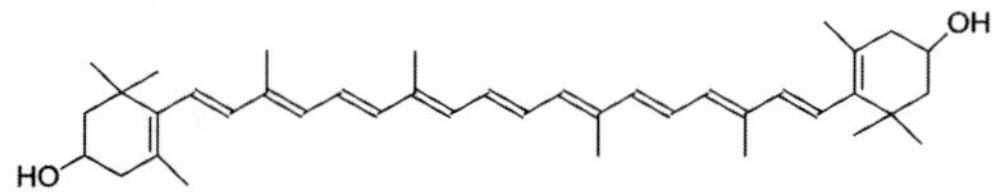<br> |

β-Ionone, 7,8-dihydro-β-ionone and megastigma-4,6,8-triene were detected prior to anthesis and their levels increased up to the time of anthesis and exhibited a fast decrease thereafter. A safranal related compound 2,2,2-trimethyl-2-cyclohexene-1,4-dione (4-oxoisophorone) increased rapidly at the anthesis stage and also in senescent stigmas, although safranal, the most typical aroma compound of the processed spice, was only detected at low levels. Like 4-oxoisophorone, acetophenone had an emission peak in the postanthesis stage.

# Biosynthesis

## Carotenoid Biosynthesis

In saffron the development of the stigma occurs concomitantly with the amyloplast-to-chromoplast transition and the stigma never turns green during this process (*33*) and parallels carotenoid accumulation (*17*).

The carotenoid precursors phytoene, phytofluene, and tetrahydrolycopene (two possible compounds 7,8,11,12-tetrahydrolycopene or ζ-carotene) and the carotenoids β-carotene, γ-carotene, lycopene and zeaxanthin have been isolated from saffron stigmas (*28*, *34*). Castillo et al. (*17*) detected also β-cryptoxanthin.

As for almost all photosynthetic organisms, in *Crocus sativus* the $C_5$ building blocks isopentenyl diphosphate (IPP) and dimethylallyl diphosphate (DMAPP) are necessary for the carotenoid biosynthesis. These blocks are supplied through the 2-C-methyl-D-erythritol 4-phosphate (MEP) pathway located in plastids. There may be a minor level of crosstalk and IPP exchange with the mevalonate (MVA) pathway in the cytosol for synthesizing the $C_5$ precursors for carotenoid formation, but carotenogenesis is not dependent on IPP import from the MVA pathway (*35*). The MEP pathway comprises a total of seven biosynthetic steps (for details the reader is referred to the recent review of Walter et al. (*36*)). After the MEP pathway, the next step in carotenoid biosynthesis is the assembly of $C_5$ units into $C_{20}$ by IPP isomerase (IDI) and GGPP (geranylgeranylpyrophosphate) synthase (GGPPS). The first committed step of carotenoid biosynthes is a head-to-head coupling of two molecules of GGPP to yield 15-*cis*-phytoene and is catalyzed by phytoene synthase (PSY). Two genes for phytoene synthase (PSY) have been identified (*37*). Castillo et al. (*17*) cloned the cDNA for PSY and observed the *CsPSY* transcript in saffron (Table I).

Phytoene desaturase (PDS) catalyzes the dual desaturation of 15-*cis*-phytoene and renders the intermediate 9,15,9′-tri-*cis*-ζ-carotene that needs to be isomerized to 9,9′-di- *cis*-ζ-carotene either by the action of light or enzymatically by Z-ISO (*38*) before two further desaturation reactions can occur (Table I). Then a second desaturase ζ-carotene desaturase (ZDS) works at the 7-8 and 7′-8′ positions. The resulting 7,9,7′,9′-tetra-*cis*-lycopene must be isomerized by another carotenoid isomerase (CrtISO) to yield all-*trans* lycopene (*39*, *40*). The most important modification of the linear $C_{40}$ backbone of all-*trans* lycopene is cyclization of terminal units to give ε- or β-rings by two competing cyclases: ε-lycopene cyclase (ε-LCY) and β-lycopene cyclase (β-LCY). When combined, these cyclases generate α-carotene, which exhibits one ε- and one β-ring, whereas the only action of β-LCY forms β-carotene. Two genes of lycopene cyclase (LCY: *CstLCYB1, CstLCYB2a*) have been identified in saffron (*41*). Hydroxylations of the ionone rings at the 3,3′-positions are performed by enzymes specific for the two types of rings (β- or ε-) with some likely crossover. The modification of the β-ionone ring towards β,β-xanthophylls is catalyzed mainly by a ferredoxin-dependent non-heme di-iron monooxygenase called BCH, whereas ε-ring hydroxylation towards the β,ε-xanthophyll lutein appears to be carried out by structurally unrelated heme-containing cytochrome P450 monooxygenases (CYP97A, CYP97C) (*42*). As for PSY and LCY, two genes of carotene hydroxylase (CHY: *CsBCH1, CsBCH2*) have been identified (Table I) (*17*).

It seems that a specific chromoplast pathway for the carotenoid biosynthesis in *Crocus sativus* generated by carotenogenic gene duplication might play an important role in the high apocarotenoid accumulation in stigmas (*43*).

Transcripts of *CsBCH1* were detected in stigmas at much higher levels than those of *CsBCH2,* and they showed a good correlation with apocarotenoid accumulation (*17*). Castillo et al. (*17*) suggested that hydroxylation of β-carotene to zeaxanthin is a rate-limiting step in saffron apocarotenoid biogenesis. Zeaxanthin plays an essential role in the xanthophyll cycle protecting the photosynthetic apparatus against high irradiance by dissipating excessive light energy into heat. β-Xanthophylls are epoxidated-de-epoxidated by zeaxanthin epoxidase (ZEP) and violaxanthin de-epoxidase (VDE) in this xanthophyll cycle. The subsequent opening of the cyclohexenyl 5-6-epoxide ring in violaxanthin gives rise to neoxanthin by the action of a neoxanthin synthase (NSY).

The expression of the genes involved in the carotenoid biosynthesis changes through the development of the stigma. Castillo et al. (*17*) reported an accumulation of zeaxanthin during the development of the stigma, accompanying the expression of *CsPSY*, phytoene desaturase (*CsPDS*), and *CsLYCb,* and the massive accumulation of *CsBCH* transcripts. The *CsPSY* and *CsPDS* reached the maximal level of expression in the orange stage and then remained constant through development of the stigma. The transcript accumulation for *CsLYCb, CsBCH2,* and *CsBCH1* reached a maximum at fully developed stigma.

Studying other crocus, Castillo et al. (*17*) observed that differences between the species in terms of relative *BCH* transcripts abundance mirrored the differences of their zeaxanthin content. They suggested that the reaction catalyzed by *CsBCH* enzyme could be the limiting step in the formation of saffron apocarotenoids in the stigma tissue.

## Biosynthesis of Carotenoid Cleavage Products. *Crocus sativus* Cleavage Dioxygenases

### Generation of Crocetin Dialdehyde and 3-Hydroxy-β-cyclocitral

In *Crocus sativus,* similarities in the stereochemical configuration of picrocrocin and zeaxanthin (*44*) and the occurrence of crocetin and related apocarotenoids in few species beyond *Crocus sativus* from the genera *Gardenia,*

*Buddleja*, *Jacquinia* and *Coleus* (*14*, *45–48*) support the apocarotenoid synthesis by cleavage of zeaxanthin followed by oxidative modifications and glycosylations rather than an enzymatic co-oxidation mechanism involving ubiquitous lipoxygenases or xanthine oxidases (*49*).

Experimental evidence for the bio-oxidative cleavage of zeaxanthin was first provided by Bouvier et al. (*49*). These authors identified and functionally characterized the *Crocus sativus* zeaxanthin cleavage dioxygenase gene (*CsZCD*). This gene coded for a chromoplast enzyme that cleaves zeaxanthin at the 7,8/7′,8′-positions and rendered crocetin dialdehyde and 3-hydroxy-β-cyclocitral *in vitro*. Its expression was restricted to the stigmas and enhanced under dehydration stress. Bouvier and coworkers also suggested a stepwise sequence with the oxidative cleavage of zeaxanthin in chromoplasts followed by the translocation of modified water-soluble derivatives into the central vacuole. Castillo et al. (*17*) reported that the differences among crocus species in the expression levels of *CsZCD* were not mirrored by their differences in the crocin content.

### Generation of β-Carotene and 4,9-Dimethyldodeca-2,4,6,8,10-pentaene-1,2-dialdehyde

More recent work has shown that *CsZCD* is a truncated *CsCCD4* sequence and that full-length *CsCCD4* catalyzes the conversion of β-carotene to β-ionone in *E. coli,* implying the 9,10(9′,10′) cleavage activity rather than a 7,8(7′,8′) one. Furthermore, no activity was found for a sequence corresponding in length to *CsZCD* in later studies (*32*, *50*). It is possible that *CsZCD* exhibited and altered cleavage specificity or that *in planta* it has additional cleavage specificities in a certain subcellular environment or guided by a specific cofactor to provide the 7,8(7′,8′) cleavage required for generation of crocetin and picrocrocin precursors. This might also happen for the rest of the enzymes (CCD1 and/or CCD4) (*36*). Therefore the catalytic activity of *CsZCD* has been proposed for reconsideration (*12*, *32*, *36*) and a specific enzyme for 7,8(7′,8′) double bonds remains to be identified.

In total, four CCD isogenes of the CCD4 and CCD1 classes (*32*) and one NCED (*51*) have been identified in *Crocus sativus*: *CsCCD4a*, *CsCCD4b*, *CsCCD1a, CsCCD1b,* and *CstNCED* (Table II). *CsCCD4a* and *CsCCD4b* have a 9,10(9′,10′) cleavage activity and catalyze the formation of β-ionone in β-carotene producing cells. *CsCCD4a* and *CsCCD4b* expression is restricted to floral tissues, and presents its highest level in the stigma. There has also been observed an inverse correlation of *CsCCD4* with carotenoid content compatible with a role in apocarotenoids formation (*32*).

*CsCCD4a* and *CsCCD4b* reached the maximum expression levels in scarlet stages when the highest levels of their reaction products were found (*18*).

In the analysis of *CCD4* genomic DNA regions in *Crocus sativus* the allele *CsCCD4a* was found with and without an intron and the allele *CsCCD4b* with a unique intron (*52*). It has been suggested that independent gain/losses have occurred when the locations of CCD4 introns within the coding region were compared with CCD4 genes from other plan species. Furthermore, when comparing the promoter region of *CsCCD4a* and *CsCCD4b* with available CCD4 gene promoters from other plant species, they highlighted the conservation of *cis*-elements involved in light response, heat stress and the absence and unique presence of *cis*-elements involved in circadian regulation and low temperature responses, respectively.

The orthologs of *AtCCD1* found in *Crocus sativus* (*CsCCD1a* and *CsCCD1b*) exhibited 9,10(9′,10′)-cleavage activity and catalyze the formation of 4,9-dimethyldodeca-2,4,6,8,10-pentaene-1,2-dialdehyde from zeaxanthin (*32*, *49*) and β-ionone from β-carotene (*32*) (cf. Table II).

*CsCCD1a* is expressed constitutively in flowers (with high levels in stigma and tepals), leaf tissues and roots (*32*, *49*). A steady expression of *CsCCD1a* throughout stigma development has been observed. However its expression in the senescent stigmas was not detected. *CsCCD1b* expression is restricted to the stigma and reached a maximum in the earlier stages and then dropped in the scarlet stages. *CsCCD1a* and *CsCCD1b* expression does not mirror the emission of apocarotenoids volatiles in the stigma. This fact could be explained by differences in subcellular location (carotenoid synthesis and accumulation take place in plastids but *CsCCD1a* and *CsCCD1b* proteins do not seem to be imported to plastids). Moreover, the lag in volatile emission might be due to the translocation of the apocarotenoids products, whose release could be regulated temporally or spatially (*32*).

Comparing both classes of CCDs, CCD4 and CCD1 differ in their subcellular location in planta being plastidial and cytosolic, respectively. The plastoglobule location of the CCD4 enzymes allows them to access to plastid carotenoids, whereas only carotenoids out of these organelles or once they have lost homeostasis or are targeted for degradation are accessible to the CCD1 enzymes. Another difference is that the CsCCD4 recombinant enzymes have higher enzyme activities and produced considerably more β-ionone than CsCCD1 (*32*).

## Generation of Crocetin Esters and Picrocrocin

In the potential generation pathway from zeaxanthin, crocetin dialdehyde would render crocetin by the action of an aldehyde oxidoreductase and UDPG-glucosyltransferases (GTASEs) would give rise to the different crocetin esters (*19*, *53*, *54*). Dufresne et al. (*53*) reported the sequential glucosylation of crocetin into its digentiobiosyl ester by two GTASE enzymes. GTASE1 catalyzes the glucose transfer on both carboxylic ends of crocetin and is responsible for the formation of monoglucosyl and diglucosyl esters. It is highly specific for crocetin and will not accept structurally similar substrates, unlike most other glucosyltransferases in plants. GTASE2 catalyzes the formation of gentiobiosyl esters. Rubio Moraga et al. (*19*) isolated a glucosyltransferase enzyme involved in crocin and crocetin glucosylation (*UGTCs2*) in *Crocus sativus* stigmas. The *in vitro* experiments with the recombinant enzyme showed the two activities describes by Dufresne et al. (*53*).

**Table II. Carotenoid Cleavage Enzymes Whose Genes Have Been Cloned and Identified in Saffron Including Their Substrates, Activities, and Products**

| *Enzyme* | *Substrate and activity* | *Products* |
|---|---|---|
| *CsZCD*[a] | 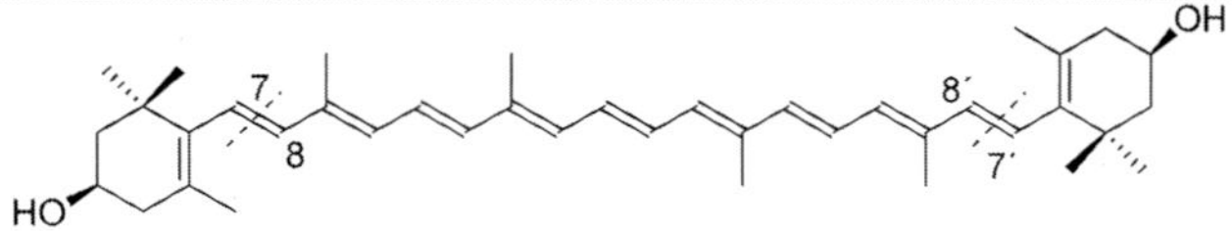<br><br>Zeaxanthin | 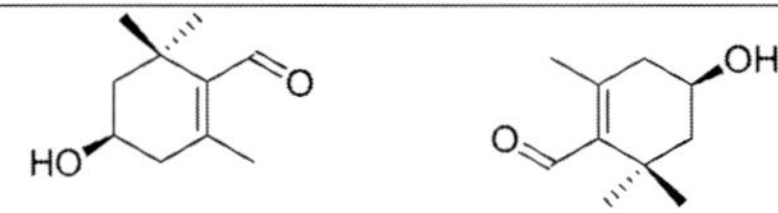<br><br>Hydroxy-safranal<br>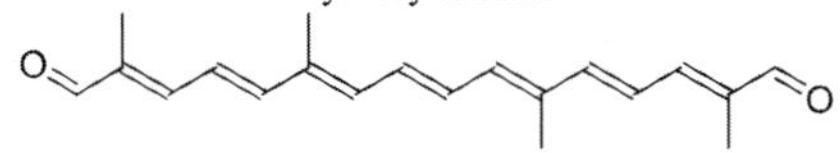<br><br>Crocetin dialdehyde |
| *CsCCD1* | 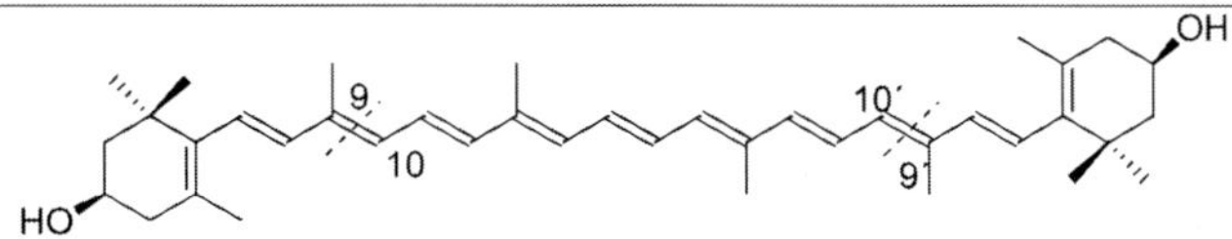<br><br>Zeaxanthin | 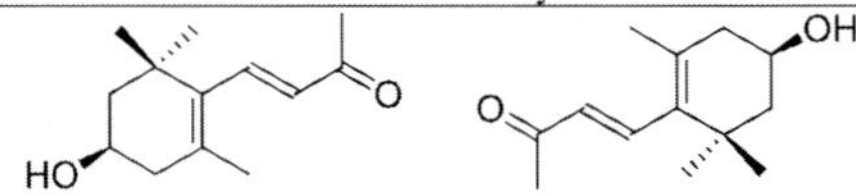<br><br>Hydroxy-β-ionone<br>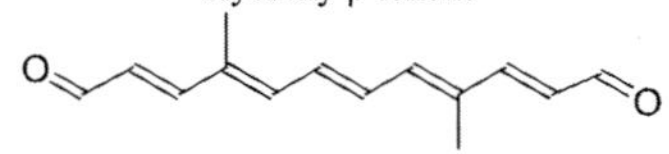<br><br>$C_{14}$ Dialdehyde |
| | 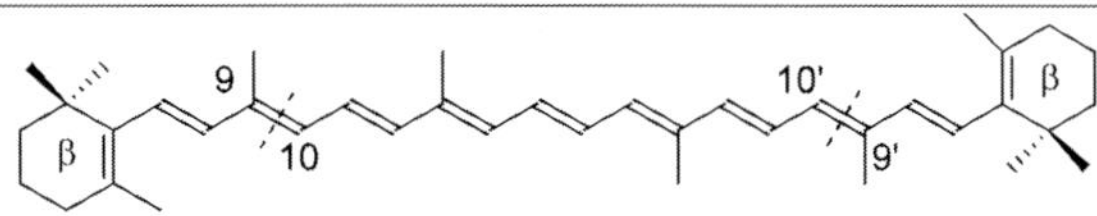<br><br>β-Carotene | 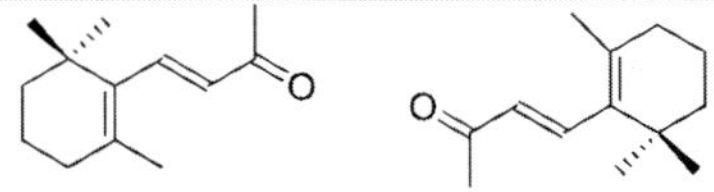<br><br>β-Ionone<br>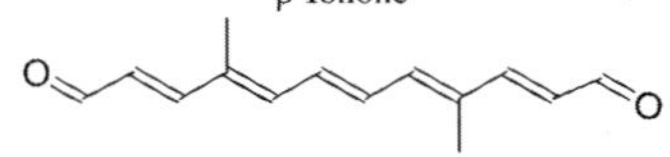<br><br>$C_{14}$ Dialdehyde |

| *Enzyme* | *Substrate and activity* | *Products* |
|---|---|---|
| *CsCCD4* | 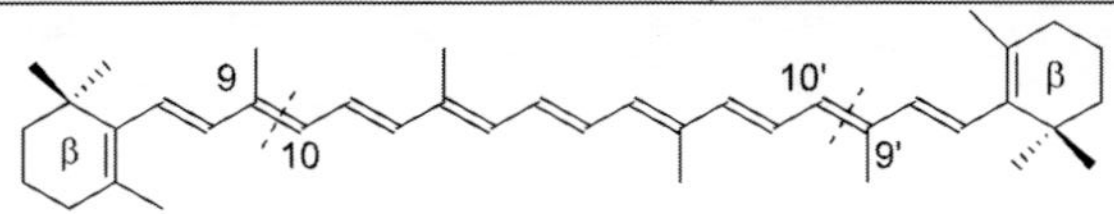<br>β-Carotene | 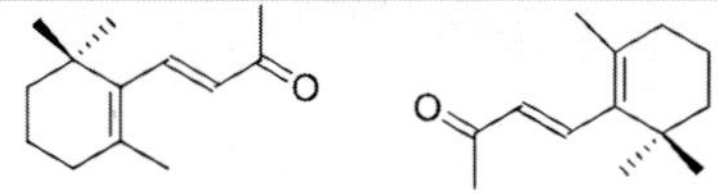<br>β-Ionone<br>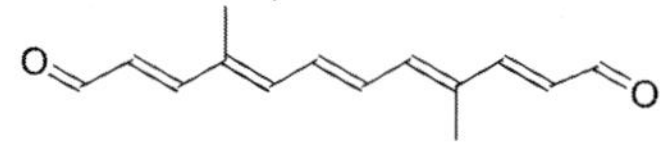<br>$C_{14}$ Dialdehyde |
| *CstNCED* | 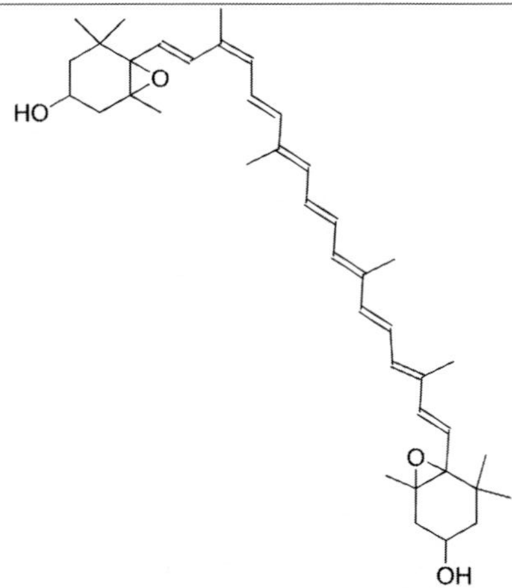<br>9′-*cis* Violaxanthin | 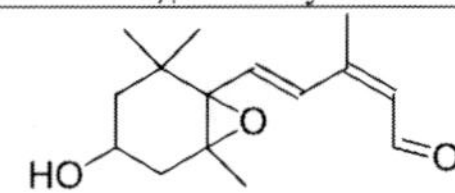<br>Xanthonin |
| | 9′-*Cis* neoxanthin | Xanthonin |

[a] Limited chracterization.

Similarly, by the action of a UDPG-glucosyltransferase 3-hydroxy-β-cyclocitral can be glucosylated to give rise to picrocrocin. Similar glucosyltransferase activities are considered be responsible for the formation of the other aroma precursors in saffron. Several EST clones from the Saffron Gene Database (*55*) encode putative glucosyltransferases of interest for determining the function and substrate specificity of these enzymes.

## Generation of Safranal and Other Aroma Compounds

The generation of safranal from picrocrocin happens either by a two-step enzymatic/dehydration process involving the intermediate 4-hydroxy-2,6,6-trimethyl-1-cyclohexen-1-carboxaldehyde (HTCC) or directly by thermal degradation (Figure 4). There is also evidence of this conversion at low pH conditions (*56*).

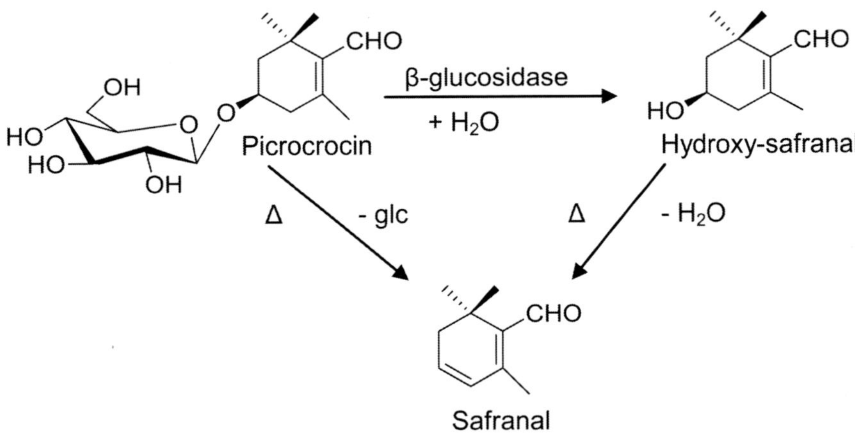


*Figure 4. Generation of safranal from picrocrocin.*

The presence of $C_8$-, $C_9$-, and $C_{13}$-compounds together with the $C_{10}$-derivatives that can be formed from picrocrocin suggested that several cleavages at different positions of the polyene chain should be involved in the generation of saffron volatiles (*57*). Carmona et al. (*58*) carried out thermal degradation studies with saffron spice. In these studies they simulated aging processes and found that while the amounts of $C_{10}$ compounds such as safranal and HTCC increased, the amounts of $C_9$ compounds such as isophorone and 2,6,6-trimethylcyclohexane-1,4-dione decreased. They tentatively identified 4,5,6,7-tetrahydro-7,7-dimethyl-5-oxo-3H-isobenzofuranone (TDOI) as a new inter-mediate generated by degradation of crocetin and crocetin esters. Figure 5 shows the hypothesis for the generation of saffron volatiles from crocetin and crocetin esters proposed by these authors.

### 9-*Cis*-Epoxycarotenoid Dioxygenases

The characterization of a 9-*cis*-epoxycarotenoid dioxygenase gene from *Crocus sativus* stigmas (*CstNCED*) showed that its expression played an important role in the regulation of abcisic acid (ABA) level during stigma senescence, corm dormancy, and drought stress (*51*). In this study a correlation between stigma senescence and the levels of *CstNCED* transcript and endogenous ABA levels was observed. Furthermore, a co-ordinated regulation of the carotenoid and ABA biosynthetic pathways in corms was suggested when the zeaxanthin levels were correlated with the increase in *CstNCED* expression.

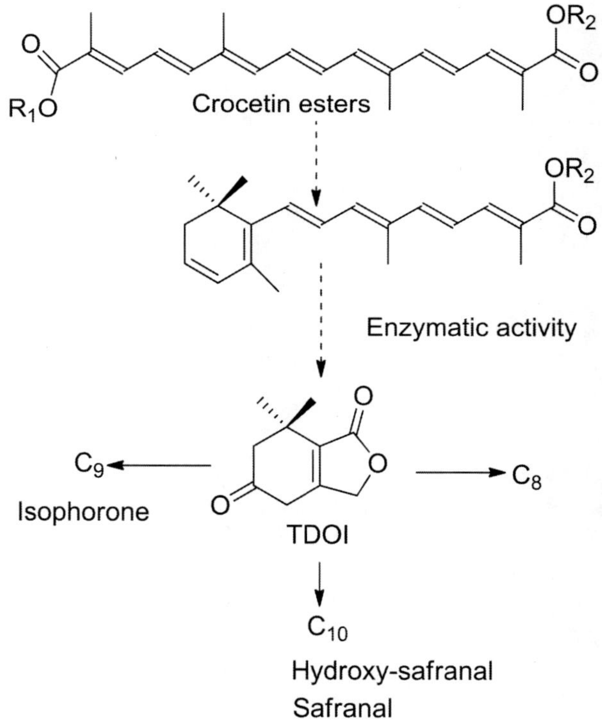


*Figure 5. Hypothesis for the generation of saffron volatiles from crocetin esters according to Carmona et al. (58); $R_1$ and $R_2$ are glucosyl, gentiobiosyl, neapolitanosyl or triglucosyl moieties.*

## Dehydration of Saffron Stigmas

There are differences in the dehydration process among the producing areas that contribute to a great extent to the quality and sensorial characteristics of saffron spice from different geographical origin (*59*). Traditional dehydration processes in some countries like India, Morocco, and Iran are carried out at room temperature directly under sunlight or in air-ventilated conditions. In other cases like in Spain, Greece, and Italy, the drying process is carried out at higher temperatures by using hot air or other heating sources (*20*). The dehydration process lets the preservation of the spice, influences its color and coloring strength (*60*); and determines saffron spice aroma through the generation of safranal and other volatile compounds (*58*).

## Importance of Carotenoid Cleavage Products

As in other plants, functions like attraction of pollinators and seed dispersers; precursors of physiologically important compounds; light harvesting pigments; photoprotection and protection against stress conditions give the carotenoids and apocarotenoids in saffron a great biological importance. Moreover, β-Ionone and other apocarotenoids exhibit antifungal and antimicrobial activity (*61*).

Importantly, the carotenoid cleavage products of saffron determined the color, flavor and aroma of saffron spice and therefore they determine its economic value. The commercial categories of the spice are determined by the main groups of the carotenoid cleavage products, especially by the crocetin esters (*62*). They are also essential in saffron authentication to distinguish genuine saffron spice from other natural colorants and/or adulterants (*22*).

Saffron spice has been used for medicinal purposes in many cultures since ancient times. Also in this field, carotenoid cleavage products have a great importance (*63*). In general, saffron is used in folk medicine as an antispasmodic, eupeptic, gingival sedative, anti-catarrhal, nerve sedative, carminative, diaphoteric, expectorant, stomachic, aphrodisiac, and anti-inflammatory remedy. Traditional knowledge regarding the medicinal effects of saffron is largely confirmed by modern pharmacological research and clinical testing. A review of the preclinical and clinical literature highlighted two major indications for saffron extracts, namely treatment of cancer and depression (*64*). In fact, pharmacological research currently focuses on antitumoural effects, whereas clinical testing has provided evidence for the traditional well-known antidepressant effects.

## References


1. Angiosperm Phylogeny Group. *Bot. J. Linn. Soc.* **2009**, *161*, 105–121.
2. Chase, M. W.; Reveal, J. L. *Bot. J. Linn. Soc.* **2009**, *161*, 122–127.
3. Strasburger, E.; Moll, F.; Schewch, H.; Schimger, A. F. N. In *Tratado de Botánica.* Marín; Barcelona, Spain, 1974.
4. Negbi, M. In *Saffron: Crocus sativus L. Medicinal and aromatic plants. Industrial profiles;* Negbi, M., Ed.; Harwood Academic Publishers: Amsterdam, Holland, 1999; pp 1–15.
5. Ghorbani, M. *World Appl. Sci. J.* **2008**, *4*, 523–527.
6. Ordoudi, S.; Tsimidou, M. In *Harvest and Quality Evaluation of Food Crop*; Dris, R., Ed.; Kluwer Academic Publishers: Dordrecht, The Netherlands, 2004; pp 215–218.

7. Carmona, M.; Zalacain, A.; Pardo, J. E.; López, E.; Alvarruiz, A.; Alonso, G. L. *J. Agric. Food Chem.* **2005**, *53*, 3974–3979.
8. Carmona, M.; Martínez, J.; Zalacain, A.; Rodríguez-Méndez, M. L.; de Saja, J. A.; Alonso, G. L. *Eur. Food Res. Technol.* **2006**, *223*, 96–101.
9. Schwartz, S. H.; Tan, B. C.; Gage, D. A.; Zeevaart, J. A. D.; McCarty, D. R. *Science* **1997**, *276*, 1872–1874.
10. Tan, B. C.; Schwartz, S. H.; Zeevaart, J. A. D.; McCarty, D. R. *Proc. Natl. Acad. Sci. U.S.A.* **1997**, *94*, 12235–12240.
11. Tan, B. C.; Joseph, L. M.; Deng, W. T.; Liu, L.; Li, Q. B. *Plant J.* **2003**, *35*, 44–56.
12. Ohmiya, A. *Plant Biotechnol.* **2009**, *26*, 351–358.
13. Speranza, G.; Dadà, G.; Manitto, P.; Monti, D.; Gramatica, P. *Gazz. Chim. Ital.* **1984**, *114*, 189–192.
14. Pfister, S.; Meyer, P.; Steck, A.; Pfander, H. *J. Agric. Food Chem.* **1996**, *44*, 2612–2615.
15. Van Calsteren, M. R.; Bissonnette, M. C.; Cormier, F.; Dufresne, Ch.; Ichi, T.; Yves Le Blanc, J. C.; Perreault, D.; Roewer, I. *J. Agric. Food Chem.* **1997**, *45*, 1055–1061.
16. Assimiadis, M. K.; Tarantilis, P. A.; Polissiou, M. G. *Appl. Spectrosc.* **1998**, *52*, 519–522.
17. Castillo, R.; Fernández, J. A.; Gómez-Gómez, L. *Plant Physiol.* **2005**, *139*, 674–689.
18. Rubio, A.; Rambla, J. L.; Ahrazem, O.; Granell, A.; Gómez-Gómez, L. *Phytochemistry* **2009**, *70*, 1009–1016.
19. Moraga, A. R.; Nohales, P. F.; Fernández-Pérez, J. A.; Gómez-Gómez, L. *Planta* **2004**, *219*, 955–966.
20. Carmona, M.; Zalacain, A.; Alonso, G. L. In *The chemical composition of saffron: color, taste and aroma*, 1st ed.; Carmona, M., Zalacain, A., Alonso, G. L., Eds.; Bomarzo: Albacete, Spain, 2006; pp 48–58.
21. Alonso, G. L.; Salinas, M. R.; Garijo, J.; Sánchez, M. A. *J. Food Qual.* **2001**, *24*, 219–233.
22. Sánchez, A. M.; Maggi, L.; Carmona, M.; Alonso, G. L. In *Progress in authentication of food and wine*; Ebeler, S., Ed.; ACS Symposium Series 1081; American Chemical Society: Washington, DC, 2012; pp 309–331.
23. Straubinger, M.; Jezussek, M.; Waibel, R.; Winterhalter, P. *J. Agric. Food Chem.* **1997**, *45*, 1678–1681.
24. Straubinger, M.; Bau, B.; Eckstein, S.; Jezussek, M.; Winterhalter, P. *J. Agric. Food Chem.* **1998**, *46*, 3238–3243.
25. Knapp, H.; Straubinger, M.; Witte, A.; Winterhalter, P. In *Frontiers of Flavor Science*; Schieberle, P., Engel, K. H., Eds.; Proceedings of the 9th Weurman Flavour Research Symposium: Garching, Germany, 2000; pp 440–444.
26. Iborra, J. L.; Castellar, M. R.; Canovas, M.; Manjón, A. *J. Food Sci.* **1992**, *57*, 714–731.
27. Sánchez, A. M.; Carmona, M.; Zalacain, A.; Carot, J. M.; Jabaloyes, J. M.; Alonso, G. L. *J. Agric. Food Chem.* **2008**, *56*, 3167–3175.
28. Kuhn, R.; Winterstein, A. *Ber. Dtsch. Chem. Ges.* **1934**, *67*, 344–347.

29. Carmona, M.; Zalacain, A.; Salinas, M. R.; Alonso, G. L. *Crit. Rev. Food Sci.* **2007**, *47*, 145–159.
30. Rödel, W.; Petrizka, M. *J. High Resolut. Chromatogr.* **1991**, *14*, 771–774.
31. Cadwallader, K. R.; Baek, H. H.; Cai, M. In *Spices: Flavor chemistry and antioxidant properties*; Riach, S. J., Ho, C. T., Eds.; ACS Symposium Series 660; American Chemical Society: Washington, DC, 1997; pp 66–79.
32. Rubio, A.; Rambla, J. L.; Santaella, M.; Gómez, M. D.; Orzaez, D.; Granell, A.; Gómez-Gómez, L. *J. Biol. Chem.* **2008**, *283*, 24816–24825.
33. Grilli-Caiola, M.; Canini, A. *Plant Biosyst.* **2004**, *138*, 43–52.
34. Pfander, H.; Schurtenberger, H. *Phytochemistry* **1982**, *21*, 1039–1042.
35. Hemmerlin, A.; Hoeffler, J. F.; Meyer, O.; Tritsch, D.; Kagan, I. A.; Grosdemange-Billiard, C.; Rohmer, M.; Bach, T. J. *J. Biol. Chem.* **2003**, *278*, 26666–26676.
36. Walter, M. H.; Strack, D. *Nat. Prod. Rep.* **2011**, *28*, 663–692.
37. Rubio, A. Ph.D. Thesis. University of Castilla-La Mancha, Albacete, Spain, 2003.
38. Chen, Y.; Li, F. Q.; Wurtzel, E. T. *Plant Physiol.* **2010**, *153*, 66–79.
39. Isaacson, T.; Ronen, G.; Zamir, D.; Hirschberg, J. *Plant Cell* **2002**, *14*, 333–342.
40. Park, H.; Kreunen, S. S.; Cuttriss, A. J.; DellaPenna, D.; Pogson, B. J. *Plant Cell* **2002**, *14*, 321–332.
41. Ahrazem, O.; Rubio-Moraga, A.; Gómez-Gómez, L. *J. Exp. Bot.* **2009**, *61*, 105–119.
42. Sandmann, G.; Romer, S.; Fraser, P. D. *Metab. Eng.* **2006**, *8*, 291–302.
43. Gómez-Gómez, L.; Rubio-Moraga, A.; Ahrazem, O. *Funct. Plant Sci. Biotechnol.* **2010**, *4*, 56–63.
44. Buchecker, R.; Eugster, C. H. *Helv. Chim. Acta* **1973**, *56*, 1121–1125.
45. Zhu, C. F.; Bai, C.; Sanahuja, G.; Yuan, W.; Farre, G.; Naqvi, S.; Shi, L. X.; Capell, T.; Christou, P. *Arch. Biochem. Biophys.* **2010**, *504*, 132–141.
46. Eugster, C. H.; Hürlimann, H.; Leuenberger, H. J. *Helv. Chim. Acta* **1969**, *52*, 89–90.
47. Tandon, J. S.; Katti, S. B.; Rüedi, P.; Eugster, C. H. *Helv. Chim. Acta* **1979**, *274*, 2706–2707.
48. Liao, Y. H.; Houghton, P. J.; Hoult, J. R. S. *J. Nat. Prod.* **1999**, *62*, 1241–1245.
49. Bouvier, F.; Suire, C.; Muttener, J.; Camara, B. *Plant Cell* **2003**, *15*, 47–62.
50. Sergeant, M. J.; Li, J.; Fox, C.; Brookbank, N.; Rea, D.; Bugg, T. D. H.; Thompson, A. J. *J. Biol. Chem.* **2009**, *284*, 5257–5264.
51. Ahrazem, O.; Rubio-Moraga, A.; Trapero, A.; Gómez-Gómez, L. *J. Exp. Bot.* **2012**, *63*, 681–694.
52. Ahrazem, O.; Trapero, A.; Gómez, M. D.; Rubio-Moraga, A.; Gómez-Gómez, L. *Genomics* **2010**, *96*, 239–250.
53. Dufresne, C.; Cormier, F.; Dorion, S. *Planta Med.* **1997**, *63*, 150–153.
54. Côté, F.; Cormier, F.; Dufresne, C.; Willemont, C. *J. Plant Physiol.* **2001**, *158*, 553–560.
55. D'Agostino, N.; Pizzichini, D.; Chiusano, M. L.; Giuliano, G. *BMC Plant Biol.* **2007**, *7*, 53.

56. Himeno, H.; Sano, K. *Agric. Biol. Chem.* **1987**, *51*, 2395–2400.
57. Winterhalter, P.; Straubinger, R. M. *Food Rev. Int.* **2000**, *16*, 39–59.
58. Carmona, M.; Zalacain, A.; Salinas, M. R.; Alonso, G. L. *J. Agric. Food Chem.* **2006**, *54*, 6825–6834.
59. Pardo, J. E.; Zalacain, A.; Carmona, M.; López, E.; Alvaruiz, A.; Alonso, G. L. *Ital. J. Food Sci.* **2002**, *4*, 413–422.
60. Carmona, M.; Zalacain, A.; Pardo, J. E.; López, E.; Alvarruiz, A.; Alonso, G. L. *J. Agric. Food Chem.* **2005**, *53*, 3974–3979.
61. Mikhlim, V. P.; Radina, A. A.; Dmitrovskii, L. P. B; Butova, L. G. *Prikl. Biokhim. Mikrobiol.* **1983**, *19*, 795–803.
62. *ISO 3632-1, 2*; ISO: Geneva, Switzerland, 2011.
63. Ulbricht, C.; Conquer, J.; Costa, D.; Hollands, W.; Iannzzi, C.; Isaac, R.; Jordan, J. K.; Ledesma, N.; Ostroff, C.; Grimes Serrano, R. I.; Shaffer, M. D.; Varghese, M. *J. Diet. Suppl.* **2011**, *8*, 58–114.
64. Schmidt, M.; Betti, G; Hensel, A. *Wien. Med. Wochenschr.* **2007**, *157* (13–14), 315–319.

## Chapter 6

# Biodegradation of Carotenoids - An Important Route to Scent Formation

**Susanne Baldermann,[1] Masaya Kato,[2] Akira Fujita,[3] Peter Fleischmann,[4] Peter Winterhalter,[4] and Naoharu Watanabe*,[1]**

**[1]Graduate School of Science and Technology, Shizuoka University, 836 Ohya, Suruga-ku, Shizuoka 422-8529, Japan**
**[2]Faculty of Agriculture, Shizuoka University, 836 Ohya, Suruga-ku, Shizuoka 422-8529, Japan**
**[3]Technical Research Center, T. Hasegawa Co. Ltd., 335 Kariyado, Nakahara-ku, Kawasaki-shi 211-0022, Japan**
**[4]Institute of Food Chemistry, Technische Universität Braunschweig, Schleinitzstrasse 20, 38106 Braunschweig, Germany**
**#Present addresses: Leibniz-Institute of Vegetable and Ornamental Crops Grossbeeren/Erfurt e.V., Grossbeeren, Germany; Institute of Nutritional Science, University of Potsdam, Nuthetal, Germany**
***E-mail: acnwata@ipc.shizuoka.ac.jp.**

$C_{13}$-apocarotenoids are potent scent compounds in flowers. Emerging areas of interest are oxidative cleavage of carotenoids and their further transformation into a wide variety of volatile apocarotenoids. Here we focus on flowers of *Osmanthus fragrans* and *Rosa chinensis* 'Mutabilis', which emit carotenoid-derived scent compounds. Enzymes involved in the selective primary cleavage reaction of carotenoids were identified and characterized. Sensory evaluation verified the importance of β-ionone in the perception of floral scent from *O. fragrans*.

## Introduction

Carotenoid derived fragrance compounds contribute to the human odor sensation originating from flowers of diverse taxa. Famous examples are roses and *Osmanthus fragrans*. Early research has led to the identification of the major scent components. In roses, however, the identification of the potent carotenoid-derived volatiles, being only present in traces, was not achieved before the 1970's. The pioneering work by Ohloff and coworkers showed that carotenoid-derived components, although present in miniscule quantities, are important contributors to the odor of rose essential oil (*1*). From that time onwards researchers focused on occurrence, importance, and formation of carotenoid-derived odors. It soon became clear that specific enzymes were involved in the formation of volatile apocarotenoids (*2*, *3*). A breakthrough was the identification of the first nine-*cis*-epoxy-dioxygenases (NCED) in maize which cleaves (9*Z*)-epoxycarotenoids specifically at the C(11,12) double bond *in planta* (*4*). Since then, the number of known genes encoding different subclasses of carotenoid cleavage dioxygenases (CCDs) has been rising steadily. CCDs are a highly heterogeneous superfamily of polyene chain oxygenases in plants as well as in bacteria, fungi, and animals. All CCDs require non-heme iron as a co-factor. In plants, nine-cis-epoxy-dioxygenases (NCEDs) involved in the formation of the abscisic acid precursor xanthoxin are considered as a special type of CCD-like enzymes. Currently four major subfamilies of plant CCDs (CCD1, CCD4, CCD7, and CCD8) are known. CCD1 and CCD4 are associated with scent and aroma formation and are involved in the production of $C_{13}$-apocarotenoids. Some CCD1 enzymes can cleave the 5,6 double bonds of lycopene to form 6-methyl-5-heten-3-one (*5*), or the 7,8 double bonds to yield geranial (*6*).

Even though the early works cited improved our understanding of apocarotenoid biosynthetic pathways of apocarotenoids, the production of the wide diversity of carotenoid-derived compounds still remains enigmatic. Various approaches are being followed to clarify modification and transformation reactions of the primarily formed volatiles (*7*).

Here, we present our recent work on the primary cleavage reaction in flowers of *O. fragrans* and the verification of the importance of β-ionone for human scent perception by sensory evaluation. Using *Rosa chinensis* 'Mutabilis' as an example, we present evidence for an enzymatic conversion of β-ionone to 7,8-dihydo-β-ionone. In this multicolored rose, β-carotene was identified as a major carotenoid and putative β-ionone precursor. Surprisingly, the major $C_{13}$-apocarotenoid was not β-ionone itself, but 7,8-dihydo-β-ionone, which prompted us to perform initial experiments to study the transformation reactions of β-ionone.

## Volatiles in Flowers of *Osmanthus fragrans*

Flowers of the evergreen Oleaceae, native to East-Asia, are characterized by a pleasant fruity-floral aroma. The color of the flowers varies between silver-white (*Osmanthus fragrans* Lour. *var. latifolius* Mak.) over gold-orange (*Osmanthus fragrans* Lour. *var. thunbergii* Mak.) to reddish (*Osmanthus fragrans* Lour. *var.*

*aurantiacus* Mak.). *Osmanthus* absolute is usually prepared from flowers of *O. fragrans* Lour. *var. thunbergii* Mak., and is used for flavoring of the most exclusive cosmetics. In this paper we describe the volatile profile of *O. fragrans* Lour. *var. aurantiacus,* the most prominent variety in Japan. To elucidate the volatile profile we applied a dynamic headspace sampling technique (*8*). We identified more than 100 volatiles, including a number of carotenoid derivatives (Figure 1). Volatiles originating from carotenoids represent more than 20% of the total volatiles in full bloom (Figure 1). Many of these compounds have low human odor perception thresholds and often contribute essentially to the odor profile of flowers, fruits, spices or processed foods.

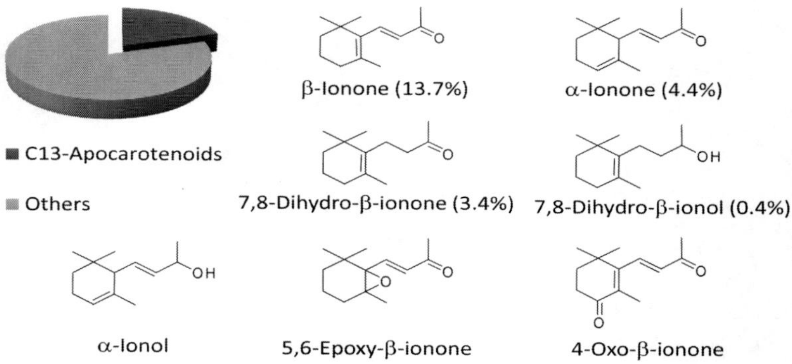


*Figure 1. Carotenoid-derived scent compounds in flowers of O. fragrans var. aurantiacus at full boom.*

## Formation of Carotenoid-Derived Volatiles in Flowers of *Osmanthus fragrans*

Carotenoids are important precursors for hormones and scent compounds in plants. Apocarotenoids are widespread and contribute to the flower odor of *Viola odorata* L. *Freesia refracta* Klatt, *Boronia megastigma* Nee, and *O. fragrans* Lour.

The dominant scent compound in flowers of *O. fragrans* has been known to be β-ionone, however, the biosynthetic pathway of the apocarotenoids in this species have remained unknown. Enzymatic carotenoid cleavage by carotenoid cleavage enzymes (CCDs) has been demonstrated in flowers of *O. fragrans*, and in other species ((*9*), *and references cited therein*). Especially the CCD subfamily 1 targets 9,10 and/or 9′,10′ double bonds of C40-carotenoids and apocarotenoids to form important volatile $C_{13}$-apocarotenoids, including the potent scent compound β-ionone. In transgenic petunia plants, a 75% decrease in β-ionone formation was observed following inhibition of *PhCCD1* gene expression (*10*).

Therefore, we targeted *OfCCD1* as a candidate gene involved in ionone biosynthesis in *O. fragrans*. We identified and characterized a CCD1 homologue based on conserved CCD sequences (*8*, *11*). By *in vivo* and *in vitro* experiments we confirmed the role of OfCCD1 in the formation of the two major scent compounds, α- and β-ionone, derived from the predominant flower pigments α- and β-carotene (*8*). *CCD1* homologues also have been identified and characterized in other species, e.g. *Petunia hybrida*, *Chrysanthemum*, or *Rosa damascena* (Figure 2).

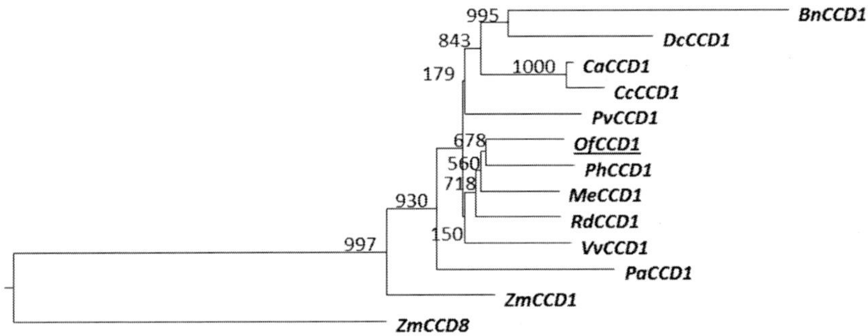


*Figure 2. Phylogenetic analysis of CCD1 nucleotide sequences from various plant species. A multiple alignment was constructed with ClustalW. Abbreviations and accession numbers are as follows: Bn Brassica napus (HQ260430), Dc Daucus carota (DQ192203), Ca Coffea arabica (DQ157179), Cc Coffea canephora (DQ157166), Pv Phaseolus vulgaris (AY029525), Of Osmanthus fragrans (AB526197), Ph Petunia hybrida (AY576003), Me Manihot esculenta (GU120077), Rd Rosa damascena (EU327776), Vv Vitis vinifera (AY856353), Pa Persea americana (AF224670.1), Zm Zea mays (DQ100346). Zea mays CCD8 (FJ957946) was included in the alignment as the outgroup. The statistical significance was evaluated by bootstrap analysis with 1000 iterative tree constructions.*

It has been demonstrated *in vitro* that these enzymes are involved in the formation of potent fragrance compounds derived from carotenoids. The release of apocarotenoids was affected by floral development and the photoperiod (*8*, *10*). These studies suggested that CCD1 enzymes probably are involved in the oxidative cleavage of carotenoids to produce the volatile scent components. However, detailed analysis of carotenoid concentration, volatile release and transcripts indicated that the activity of CCD1 enzymes is not sufficient to account for the total emission of C13-apocarotenoids (*8*). Additional work is needed to clarify the contribution of other carotenoid cleavage enzyme or non-enzymatic reactions. CCD4 enzymes from *Crocus sativus, Rosa damascena, Osmanthus fragrans, Malus domestica*, and *Chrysanthemum morifolium* (*12–14*), which can cleave carotenoids and apocarotenoids with varying degrees of efficiency, could make a contribution. A link between CCD4 transcript levels and the color of flowers has been demonstrated (*details are given in this book in the chapter "Involvement of CCD4 in Determining Petal Color" by A. Ohmiya*). A comprehensive analysis of the contribution of various CCD subclasses to carotenoid accumulation and scent formation might be helpful to gain further insights into carotenoid metabolism.

## Stereoselectivity of β-Ionone Formation

Non-symmetric carotenoids such as α-carotene with one ε- and one β-ring end group exist as 6′*R*- or 6′*S*-enantiomers. The 6′*S*-enantiomer of α-carotene has not been reported under natural conditions (*15*). Using authentic reference material, we identified 6′*R*-α-carotene by chiral HPLC analysis as the natural enantiomer in flowers of *O. fragrans*. This was a relevant finding, because *R*- and *S*-*α*-ionone, the putative cleavage products of OfCCD1, exhibit different aroma characteristics. *R*-*α*-ionone combines unique and strong violet-like, fruit-like, and raspberry-like aromas, whereas the flavor description for *S*-α-ionone varies from fresh juicy greenish, wood-like, cedarwood-like, raspberry-like, to β-ionone-like (*16*). Moreover, the floral odor threshold of *R*-α-ionone is lower (0.5 ppb) compared to its *S*-enantiomer (20–40 ppb) (*17*). Therefore, it was of special interest to determine whether the selective enzymatic cleavage of α-carotene by OfCCD1 is an enantioselective reaction. The enzymatic cleavage reaction with the purified recombinant OfCCD1 enzyme (*8*, *11*) yielded solely *R*-α-ionone (Figure 3). This indicated that an enantioselective enzymatic oxidation of carotenoids takes places *in planta* which contributes to the natural occurrence of *R*-α-ionone in flowers of *O. fragrans*. It would be interesting to study the stereochemistry of the reaction with other racemic substrates.

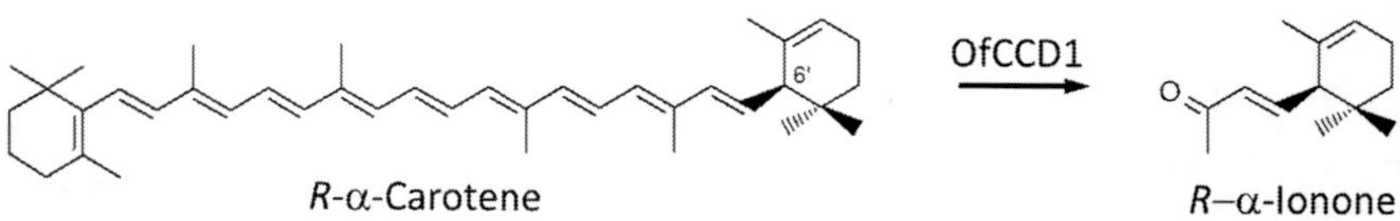


*Figure 3. Enantio-selective cleavage by OfCCD1 of R-α-carotene to R-α-ionone occurs in flowers of O. fragrans.*

## Sensory Evaluation of *O. fragrans* Essential Oil

*Osmanthus fragrans* is a famous traditional flower in East Asia and is widely cultivated as an ornamental plant today. The interest in the unique fragrant flowers is long-standing, and important odor constituents have been identified (*18*). Moreover, it has been demonstrated that the essential oil composition and scent release change over floral development and are affected by the photoperiod (*8*, *19*). Some carotenoid-derived volatiles are characterized by very low human odor perception thresholds and are among the most potent scent components. Thus we were interested in their contribution to the floral scent of *O. fragrans*. We extracted the essential oil of flowers in full blossom, which release the highest quantities of fragrant components (*8*), and subjected the extract to aroma-extract dilution analysis (AEDA). AEDA was used to identify the volatiles with the highest odor potency. The extract and the stepwise diluted samples (ranging from 4-262144) were subjected to gas chromatography–olfactometry and evaluated by three individuals. Table 1 summarizes the most potent constituents and their flavor description noticed at a dilution of 256 or above. In addition to β-ionone, γ-decalactone, linalool, vanillin, and methional have been identified as characteristic compounds in *O. fragrans* flowers. In contrast to γ-decalactone and linalool, the perception thresholds for vanillin, methional, and β-ionone differed between the panelists. The differences in the sensory perception of β-ionone covered three orders of magnitude. This indicates an enormous variation in the individual odor perception threshold for carotenoid-derived volatiles, which has not yet been adequately understood. Similar differences have been found also for α-ionone (*17*).

**Table 1. Aroma-Extract Dilution Analysis (AEDA) of *O. fragrans* Essential Oil. Listed Are Components Recognized at a Dilution of 256 or Above**

| *Compound name* | *Dilution factor* | *Odor description* |
|---|---|---|
| β-Ionone | 256-262144 | floral, sour, powdery, fresh green seaweed |
| γ-Decalactone | 65536 | peach like |
| Linalool | 4096 | floral, linalool-like |
| Vanillin | < 256-1024 | sweet |
| Methional | < 256-256 | potato-like |

## Rosa chinensis 'Mutabilis'

*Rosa* species are another famous example of flowers in which carotenoid-derived volatiles prominently contribute to floral scent. On the basis of the determination of floral pigments and of $C_{13}$-apocarotenoids, the presence of specific 9,10 (9′,10′) carotenoid cleavage enzymes in roses has been postulated (*3*). Twenty years passed from the demonstration of the initial evidence (*20*) to the confirmation of the presence and functional characterization of CCD1 and CCD4 in *R. damascena* Mill. (*21*). The recombinant RdCCD1 enzyme cleaves a multitude of substrates *in vivo* and *in vitro*, while RdCCD4 preferentially acts on apocarotenoids (*13, 21*). Using crude enzyme extracts of *R. chinensis* 'Mutabilis', we verified the occurrence of cleavage reactions in carotenoids and apocarotenoids in these flowers (*22*). However, only 7,8-dihydro-β-ionone was emitted, and we postulated that 7,8-dihydro-β-ionone is derived from β-ionone (Figure 4). To test our hypothesis, we administered deuterium labelled β-ionone to flowers. After an incubation period of 24 hours, the presence of deuterium-labelled 7,8-dihydro-β-ionone was established. We observed an increase of three mass units and thus confirmed β-ionone as a precursor of 7,8-dihydro-β-ionone. A chemical transformation from β-ionone to 7,8-dihydro-β-ionone does not take place (*22*), and it would be very interesting to identify and characterize the catalyzing enzyme.

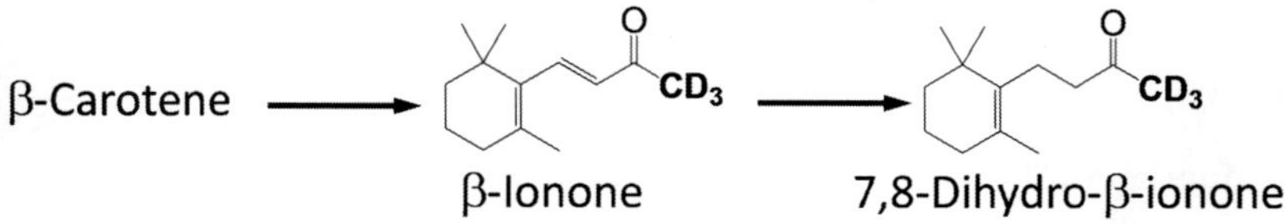


*Figure 4. Proposed biosynthesis of 7,8-dihydro-β-ionone in flowers of R. chinensis 'Mutabilis'. First, β-ionone is formed by the cleavage of β-carotene by a CCD1-like enzyme. Second, β-ionone is enzymatically converted to 7,8-dihydro-β-ionone. The pathway was elucidated using stable isotope labelled [$^2H_3$]β-ionone.*

## Future Perspective

With the identification and characterization of CCD1, an important step in the formation of carotenoid-derived volatiles has been clarified. However, our present knowledge is not sufficient and further research is required to understand the complex pathways that yield the diverse apocarotenoids found in flowers, fruits, vegetables, and processed food.

## References

1. Ohloff, G. *Riechstoffe und Geruchssinn – Die molekulare Welt der Düfte*; Springer-Verlag: Berlin, Germany, 1990; pp 152–156.
2. Enzell, C. *Pure Appl. Chem.* **1985**, *57*, 693–700.
3. Eugster, C.-H.; Märki-Fischer, E. *Angew. Chem., Int. Ed. Engl.* **1991**, *30*, 654–672.
4. Schwartz, S.-H.; Tan, B.-C.; Gage, D.-A.; Zeevaart, J.-A.-D.; McCarty, D. R. *Science* **1997**, *276*, 1872–1874.
5. Vogel, J.-T.; Tan, B.-C.; McCarty, D. R.; Klee, H.-J. *J. Biol. Chem.* **2008**, *283*, 11364–11373.
6. Ilg, A.; Beyer, P.; Al-Babili, S. *FEBS J.* **2009**, *3*, 736–747.
7. Mathieu, S.; Wirth, J.; Sauvage, F.-X.; Lepoutre, J.-P.; Baumes, R.; Günata, Z. *Plant Cell, Tissue Organ. Cult.* **2009**, *97*, 203–213.
8. Baldermann, S.; Kato, M.; Kurosawa, M.; Kurobayashi, Y.; Fujita, A.; Fleischmann, P.; Watanabe, N. *J. Exp. Bot.* **2010**, *61*, 2967–2977.
9. Walter, M.-H; Strack, D. *Nat. Prod. Rep.* **2011**, *28*, 663–692.
10. Simkin, A.-J.; Underwood, B.-A.; Auldridge, M.; Loucas, H. M.; Shibuya, K.; Schmelz, E.; Clark, D. G.; Klee, H. J. *Plant Physiol.* **2004**, *136*, 3504–3514.
11. Baldermann, S.; Kato, M.; Fleischmann, P.; Watanabe, N. *Acta Biochim. Pol.* **2012**, *59*, 79–81.

12. Rubio, A.; Rambla, J.-L.; Santaella, M.; Gomez, M. D.; Orzaez, D.; Granell, A.; Gomez-Gomez, L. *J. Biol. Chem.* **2008**, *283*, 24816–24825.
13. Huang, F.-C.; Molnár, P.; Schwab, W. *J. Exp. Bot.* **2009**, *60*, 3011–3022.
14. Ohmiya, A.; Kishimoto, S.; Aida, R.; Yoshioka, S.; Sumimoto, K. *Plant Physiol.* **2006**, *142*, 1193–1201.
15. Britton, G.; Liaaen-Jensen, S.; Pfander, H. *Carotenoids- Handbook*; Birkhäuser-Verlag: Basel, Switzerland, 2004; p 7.
16. Yamamoto, T.; Yagi, K.; Ishida, K. U. S. Patent Application 20090216039, 2009.
17. Werkhoff, P.; Bretschneider, W.; Guentert, M.; Hopp, R.; Surburg, H. In *Flavour Science and Technology*; Bessiere, Y., Thomas, A.-F., Eds; J. Wiley & Sons: New York, 1990; pp 33–36.
18. Ishiguro, T.; Koga, N.; Nara, K. *Yakugaku Zasshi* **1957**, *77*, 566–567.
19. Wang, L.; Li, M.; Jin, W.; Li, S.; Zhang, S.; Yu, L. *Food Chem.* **2009**, *114*, 233–236.
20. Suzuki, M.; Matsumoto, S.; Fleischmann, H.-P.; Shimada, H.; Yamano, Y.; Ito, M.; Watanabe, N. In *Carotenoid-derived Aroma Compounds*; Winterhalter, P., Rouseff, R.-L., Eds; ACS Symposium Series 802; American Chemical Society: Washington DC, 2002; pp 89–101.
21. Huang, F.-C.; Horváth, G.; Molnár, P.; Turcsi, E.; Deli, J.; Schrader, J.; Sandmann, G.; Schmidt, H.; Schwab, W. *Phytochemistry* **2009**, *70*, 457–464.
22. Baldermann, S.; Hirata, H.; Ueda, Y.; Winterhalter, P.; Fleischmann, P.; Watanabe, N. In *Advances and Challenges in Flavor Chemistry and Biology*; Hofmann, T., Meyerhof, W., Schieberle, P., Eds.; DFA: Freising, Germany, 2010; pp 201–206.

## Chapter 7

# $C_{13}$-Apocarotenoids: More than Flavor Compounds?

**Susanne Baldermann,[*,1,4] Masayoshi Yamamoto,[1] Ziyin Yang,[1] Tatsuya Kawahashi,[2] Kazuyoshi Kuwano,[3] and Naoharu Watanabe[1]**

**[1]Graduate School of Science and Technology, Shizuoka University, 836 Ohya, Suruga-ku, Shizuoka 422-8529, Japan**
**[2]Faculty of Agriculture, Shizuoka University, 836 Ohya, Suruga-ku, Shizuoka 422-8529, Japan**
**[3]Graduate School of Science and Technology, Nagasaki University, 1-14 Bunkyo-machi, Nagasaki 852-8521, Japan**
**[4]Present addresses: Leibniz-Institute of Vegetable and Ornamental Crops Grossbeeren/Erfurt e.V., Theodor-Echtereyerweg 1, 14979 Grossbeeren, Germany; Institute of Nutritional Science, University of Potsdam, Arthur-Scheunert-Allee 114-116, 14558 Nuthetal, Germany**
**[*]E-mail: baldermann@igzev.de.**

Carotenoid-derived volatiles are among the most potent flavor components, which essentially contribute to the flavor of fruits, vegetables and processed foods. These compounds are also released by diverse algae taxa and influence aquatic odors. Here, we will present our latest findings about the biosynthesis of $C_{13}$-apocarotenoids in green macro-algae and their biological activities. It is well established that some $C_{13}$- apocarotenoids are among the most potent flavor compounds, however little is known about their biological functions. Recent studies indicate that some carotenoid derived breakdown products exhibit unknown functions *in planta*. This presentation will show that $C_{13}$-apocarotenoids or their metabolites exhibit growth regulating properties. We observed first evidence of growth regulating effects of various $C_{13}$-apocarotenoids with two species, namely *Enteromorpha compressa* L. Ness and *Lemna pausicostata* L. $C_{13}$-apo-carotenoid treated and non-treated samples differed in their growth characteristics and could be distinguished by non-targeted analysis of their non-volatile metabolome.

Metabolites derived from carotenoids can function as growth factors, hormones, signaling or aroma-active constituents (*1*). Volatile apocarotenoids, such as the potent flavor compound β-ionone, are produced by a diversity of algae taxa, for example *Ulothrix fimbriata* (*2*) and Asakusa Nori (*Poryphyra tenera*) (*3*). However, little is known about their biosynthesis and functions in seaweed. In the last decade, there has been special interest in the isolation and characterization of carotenoid cleavage dioxygenases (CCDs) involved in the formation of apocarotenoids. These enzymes selectively cleave double bonds of the carotenoid backbone and can utilize a broad range of substrates.

Recently, we demonstrated that functional enzymes can be also found in the multi-cellular green algae *E. compressa*, cultivated and used as seasoning in Japan (*4*). Beyond their importance as aroma or scent compounds, it was of special interest to elucidate further biological functions of apocarotenoids. They can *inter alia* function as hormones, attractants for pollinators, as allelochemicals and so forth (cf. Table 1).

**Table 1. Examples of Biological Functions of Short-Chain Apocarotenoids Apart from Aroma and Hormonal Properties**

| *Function* | *Reference* |
|---|---|
| Allelochemicals | (*5–9*) |
| Pheromones | (*10*) |
| Effect on carotenoid biosynthesis | (*11*) |
| Effect on glucose uptake | (*11*) |
| Repellent for freshwater nematodes | (*12*) |
| Food finding cues for freshwater herbivores | (*13*) |

We were interested, if the endogenous volatile apocarotenoids found in *E. compressa* exhibit further functions in the aquatic environment and established a laboratory culturing system (*14*) to study the effects of apocarotenoid treatment on this green algae. Interestingly, the supplementation of apocarotenoids positively affected the longitudinal growth. To prove if our findings are solely related to the growth of algae or also applicable for other species we carried out a series of experiments with the model plant *Lemna pausicostata* L. In this species only low apocarotenoid concentrations increased the number of fronds, whereas at higher amounts the growth was inhibited.

## Major Endogenous Apocarotenoids in *E. compressa*

A number of carotenoid derived volatiles have been detected in various algae. Ohloff and coworkers reported numerous carotenoid-derived volatiles in macro-algae (*3*). In Asakusa-nori (*Porphyra sp.*), a red algae frequently consumed in

Japan, numerous $C_9$-, $C_{11}$-, and $C_{13}$-apocarotenoids have been identified. In the present study we could identify 23 major volatile compounds including β-ionone and 7,8-dihydro-β-ionone as abundant $C_{13}$-apocarotenoids (Figure 1). β-Ionone has an extremely low perception threshold and belongs to the most potent flavor molecules derived from carotenoids and contributes to the aroma of algae and the aquatic odor.

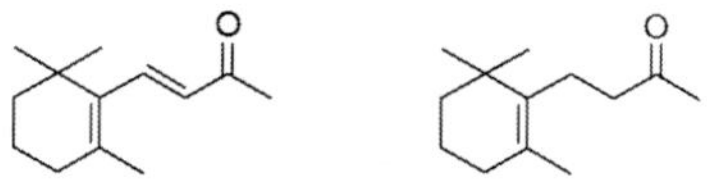


*Figure 1. Major endogenous $C_{13}$-apocarotenoids β-ionone and 7,8-dihydro-β-ionone in E. compressa.*

## Formation of Apocarotenoids in Algae

Carotenoid cleavage dioxygenases (CCDs) are widely distributed and a gene encoding a CCD1 was first identified in cyanobacteria (*Nostoc* sp. PCC 7120) by Marasco and coworkers (*15*). This enzyme catalyses the cleavage of carotenoids and long-chain non-volatile apocarotenoids. Meanwhile numerous genes encoding CCDs in cyanobacteria are known (*16*). Using a biochemical approach we identified CCD-like enzymes in *E. compressa*. Like other CCDs (CCD1 and CCD4) identified and characterized *in planta*, the algae derived enzyme can utilize carotenoids and apocarotenoids as substrate giving rise to a formation of β-ionone (*4*). The biosynthesis of the abundant compound 7,8-dihydro-β-ionone most likely takes places in two steps: (I) cleavage of β-carotene to β-ionone, (II) reduction of β-ionone to 7,8-dihydro-β-ionone by specific enzymes as will be discussed in the chapter "Occurrence, formation, and importance of volatile apocarotenoids in selected flowers" of this book.

## Apocarotenoids as Allelochemicals

Secondary metabolites with allelopathic activity are present in plants, algae, bacteria, and fungi (*5–9*). Allelopathic substances are produced by organisms to influence growth, survival, and reproduction of other organisms. Plants as well as algae are able to secrete a wide range of compounds into the environment. In the rhizosphere these components can contribute to regulate the growth of the soil bacteria community. Several studies demonstrated that carotenoid derived compounds are bioactive allelochemicals (Table 1). For example eight bioactive apocarotenoids have been isolated from sunflowers (*7, 8*), which mainly showed inhibitory effects on the growth of wheat coleoptiles. Recently, 3-hydroxy-β-ionone was identified as allelochemical released by *Rhynchosstegium pallidifolium*. The methanolic extract of *R. pallidifolium* inhibited the root growth of cress, lettuce, alfalfa, and ryegrass (*9*).

## Effects of Apocarotenoid Treatment on the Growth of *E. compressa*

A study investigating the susceptibility to blue mold of tobacco plants showed that besides changed disease resistance, ionone-treated plants showed growth and morphological changes (*17*). The plant height increased by an average of 20-30%, flowering was accelerated, and apical dominance was weakened. Moreover, the knowledge about carotenoid-derived plant hormones (abscisic acid and strigolactones) is not sufficient enough to explain all experimental results in relation to growth and a bypass regulating root and shoot development remains unknown (*18*). Interestingly, current research has shown that β-ionone also acts as stress signal that mediates gene response to singlet oxygen in plants (*19*).

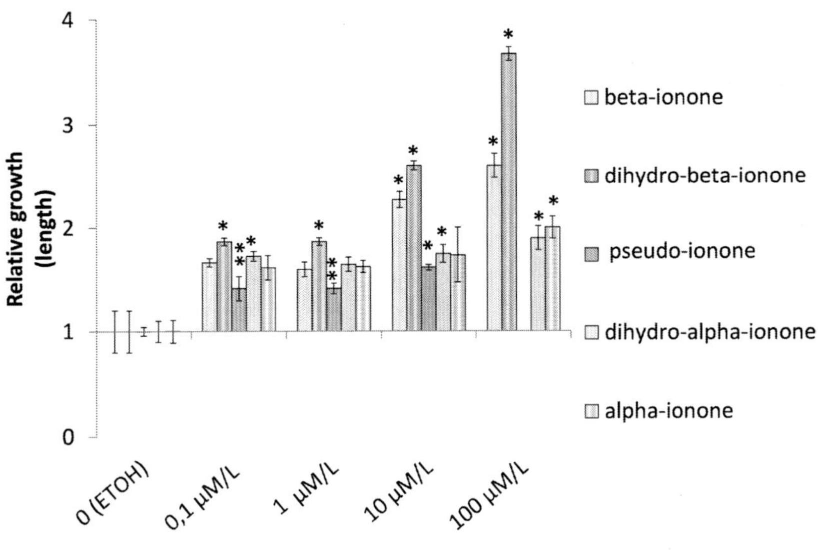


*Figure 2. Relative changes in longitudinal growth of E. compressa after treatment with selected C13-apocarotenoids. The changes are expressed as mean ± SD from independent triplicates relative to the control group. Statistical analysis was performed using ANOVA and the Tukey's test was applied to compare the means (*p < 0.05, ** p < 0.1).*

We were interested if the two endogenous $C_{13}$-apocarotenoids, β-ionone and 7,8-dihydro-β-ionone, affect the growth of *E. compressa*. To test the biological activity, these two components were added to the culture media in a range of 0-100 µM/L and the changes in growth were evaluated based on the length. In addition the biological effects of addition of α-ionone, 7,8-dihydro-α-ionone, and pseudoionone were determined. At all concentrations the two endogenous $C_{13}$-apocarotenoids, β-ionone and 7,8-dihydro-β-ionone, showed a growth stimulating effect in a concentration dependent manner. Higher apocarotenoid concentrations yielded a higher length. In case of 7,8-dihydro-β-ionone even stronger effects were observed. After treatment the length of the algae increased by a factor of 3.7 compared to the control group. β-Ionone treatment at equal concentration was less stimulating and resulted in a 2.6 fold increase in length. α-Ionone and 7,8-dihydro-α-ionone treatment showed similar results to β-ionone and the lengths increased in the samples treated with 0.1 μM/L and 1 μM/L apocarotenoid by a factor of 1.6. Similarly, the growth was stimulated by α-ionones at concentrations of 10 μM/L or 100 μM/L. At higher concentrations exogenously applied β-ionone and 7,8-dihydro-β-ionone more efficiently increased the growth (Figure 2).

Whether the observed effects are due to the compounds themselves, their metabolites or some constituents produced in response to apocarotenoid treatment still remains unknown.

In order to identify norisoprenoid induced changes in the pattern of secondary metabolites of *E. compressa* we applied non-targeted metabolite analysis by UPLC-ToF-MS. The results of the statistical analysis of the data obtained in negative ionization mode are summarized in Figure 3. We focused on the variance of PC1 and PC2, which accumulative accounts for 46% of the total variance of the data set. As shown in Figure 3, a clear classification between the treated groups and untreated group was obtained. The relatively low accumulative variance of PC1 and PC2 is the result of the high data noise level within the groups. In order to minimize this error the number of samples should be increased in future studies.

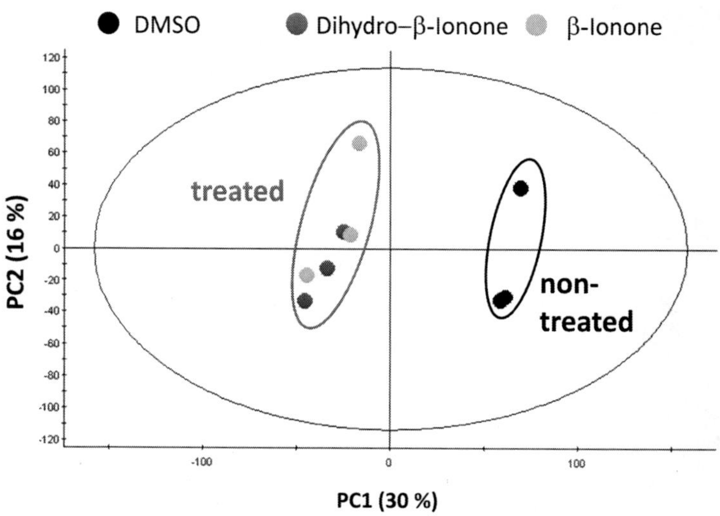


*Figure 3. PCA of metabolites ([M-H]⁻ m/z 70–1000) of treated (β-ionone, 7,8-dihydro-β-ionone) and non-treated (DMSO, control) methanolic extracts (70% methanol) of E. compressa cultures. The PCA was performed using a combination of m/z and retention times of sample chromatograms by MarkerLynx, an application manager of MassLynx Version 4.1. The experiments were performed as independent triplicates.*

Based on the molecular masses we could demonstrate that membrane components are subjected to changes (Table 2). Sulfolipids, found in diverse algae taxa and *in planta*, exhibit various functions. For example they are involved in plastid development and assembly of the photosynthetic apparatus. Moreover, changes in response to plant stress and herbivores have been reported.

**Table 2. Compounds preliminary identified based on UPLC-ToF-MS⁻ analysis (70% methanolic extract)**

| *[M-H]⁻* | *Fomular* | *Compound name* | *Change* | *Reference* |
|---|---|---|---|---|
| 277.18 | $C_{17}H_{26}O_3$-H | fatty acid derivative | ↑ | |
| 555.28 | $C_{25}H_{47}O_{11}S$-H | SQMG (C16) | ↓ | |
| 699.38 | $C_{35}H_{60}O_7P_2$+FA-H | heptaprenyl diphosphate | ↓ | |
| 787.46 | $C_{42}H_{71}O_{10}P$+Na-2H | phosphatidylglycerol | ↑ | (*20–25*) |
| 813.48 | $C_{43}H_{73}O_{12}S$-H | SQDG (C16:1, 18:3 or C16:3, C18:1) | ↑ | |
| 815.49 | $C_{43}H_{75}O_{12}S$-H | SQDG (C16, C18:3 or C16:3,18:0) | ↑ | |

FA formic acid, SQMG sulfoquinovosyl monoacylglycerol, SQDG sulfoquino-vosyl diacylglycerol.

## Effects of Apocarotenoid Treatment on the Growth of *L. pausicostata* L.

In order to check if apocarotenoid treatment related changes in growth are restricted to algae we additionally tested the biological functions of these components with *Lemna pausicostata* L.; a well-established model for studies of plant growth factors. We tested the identical apocarotenoids, namely β-ionone, 7,8-dihydro-β-ionone, α-ionone, 7,8-dihydro-α-ionone, and pseudoionone at equal concentrations. Relative changes in growth where evaluated based on the number of fronds in relation to the control groups (Figure 4). Apocarotenoids in concentrations of 0.1 μM/L and 1 μM/L increased the number of fronds, whereas the strongest effect was observed for 7,8-dihydro-β-ionone at a concentration of 1 μM/L. At concentrations of 10 μM/L we observed a diverse response, though not statistical significant at a probability level of $p < 0.05$. Stimulating effects were found for 7,8-dihydro-β-ionone, 7,8-dihydro-α-ionone, and α-ionone. β-Ionone treatment had no effect, whereas pseudoionone slightly reduced the growth. The growth of *L. pausicostata* was strongly inhibited by all tested apocarotenoids at 100 µM/L. A small selection of volatiles derived from other biosynthetic pathways was tested at a concentration of 0.1 µM/L as well. Hexenol as fatty acid derived volatile, 2-phenylethanol as representative of the shikimate pathway and linalool as common terpenoid resulted in relative changes of growth of maximal 0.1 fold (data not shown). These results indicate that growth stimulating effects might be limited to the class of volatile apocarotenoids.

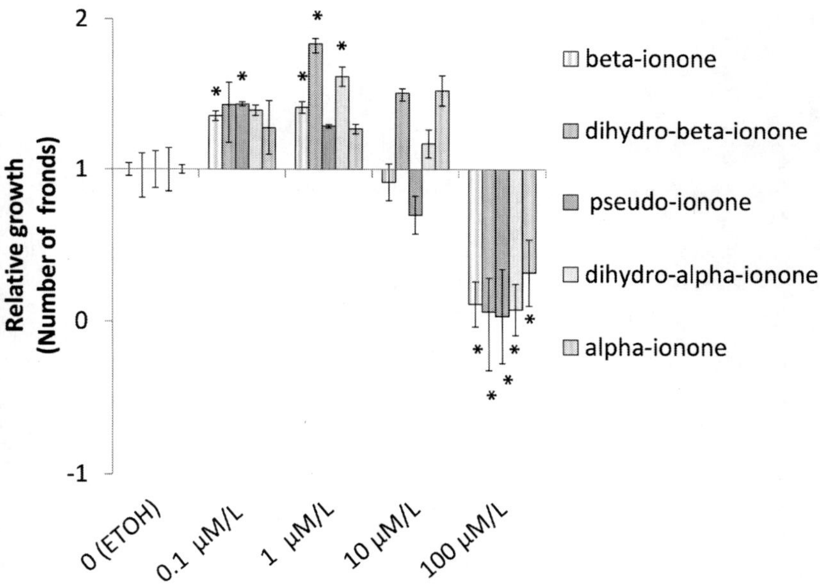


*Figure 4. Relative changes in growth of L. pausicostata after treatment with selected $C_{13}$-apocarotenoids. The changes are expressed as mean ± $S_D$ from independent triplicates relative to the control group. Statistical analysis was performed using ANOVA and the Tukey's test was applied to compare the means (*$p < 0.05$).*

## Future Perspective

This study suggests that carotenoid derived volatiles are more than flavor components and act themselves, as resulting metabolites or signaling molecules as growth regulating components. Apart from their diverse functions as scent components, repellent against herbivores, attractants for pollinators, allelochemicals, they might act as so far unknown signaling molecules. Subsequent systematic studies with model plants will provide further insights into yet unknown biological activities of apocarotenoids.

## References


1. Ohmiya, A. *Plant Biotechnol.* **2009**, *26*, 351–358.
2. Fink, P.; Elert, E.; Jüttner, F. *J. Chem. Ecol.* **2006**, *32*, 1867–1881.
3. Flament, I.; Ohloff G. In *Progress in Flavour Research*; Adda, J., Ed.; Proceedings of the 4th Weurman Flavour Research Symposium; Elsevier: Amsterdam, 1985; pp 281–300.
4. Baldermann, S.; Mulyadi, A. N.; Yang, Z. Y.; Murata, A.; Fleischmann, P.; Winterhalter, P.; Knight, M.; Finn, T. M.; Watanabe, N. *J. Sep. Sci.* **2011**, *34*, 2759–2764.
5. Jüttner, F. *Z. Naturforsch.* **1979**, *34c*, 186–191.
6. Dietz, H.; Winterhalter, P. *Phytochemistry* **1996**, *42*, 1005–1010.
7. Macías, F. A.; Varela, R. M.; Torres, A.; Oliva, R. M.; Molinillo, J. M. G. *Phytochemistry* **1998**, *48*, 631–636.
8. Macías, F. A.; López, A.; Varela, R. M.; Torres, A.; Molinillo, J. M. G. *Phytochemistry* **2004**, *65*, 3057–3063.
9. Kato-Noguchi, H.; Seki, T.; Shigemori, H. *J. Plant Physiol.* **2010**, *167*, 468–471.
10. *The Pherobase: Database of pheromones and semiochemicals*; http://www.pherobase.com (accessed July 2012).
11. Watson, S. B. *Phycologia* **2003**, *42*, 332–350.
12. Höckelmann, C.; Moens, T.; Jüttner, F. *Limnol. Oceanogr.* **2004**, *49*, 1809–1819.
13. Fink, P.; von Elert, E.; Jüttner, F. *J. Chem. Ecol.* **2006**, *32*, 1867–1881.
14. Kuwano, K.; Sakurai, R.; Motozu, Y.; Kitade, Y.; Saga, N. *J. Phycol.* **2008**, *44*, 364–373.
15. Marasco, E. K.; Vay, K.; Schmidt-Dannert, C. *J. Biol. Chem.* **2006**, *281*, 31583–31593.
16. Cui, H.; Wang, Y.; Qin, S. *Comp. Funct. Genomics* 2012, Article ID 164690, doi:10.1155/2012/164690.
17. Salt, S. D.; Tuzun, S.; Kuć, J. *Physiol. Mol. Plant Pathol.* **1986**, *28*, 287–297.
18. Cazzonelli, C. I.; Yin, K.; Pogson, B. J. *Plant Signaling Behav.* **2009**, *4*, 339–341.
19. Ramel, F.; Birtic, S.; Ginies, C.; Soubigou-Taconnat, L.; Triantaphylidès, C.; Havaux, M. *Proc. Natl. Acad. Sci. U.S.A* **2012**, *109*, 5535–5540.
20. Mohn, T.; Plitzko, I.; Hamburger, M. *Phytochemistry* **2009**, *70*, 924–934.
21. Scoparo, C. T.; de Souza, L. M.; Dartora, N.; Sassaki, G. L.; Gorin, P. A. J.; Iacomini, M. *J. Chromatogr., A.* **2012**, *1222*, 29–37.
22. Devaiah, S. P.; Roth, M. R.; Baughman, E.; Li, M.; Tamura, P.; Jeannotte, R.; Welti, R.; Wang, X. *Phytochemistry* **2006**, *67*, 1907–1924.
23. Dembitsky, V. M.; Pechenkina-Shubina, E. E.; Rozentsvet, O. A. *Phytochemistry* **1991**, *30*, 2279–2283.
24. Xu, J.; Chen, D.; Yan, X.; Chen, J.; Zhou, C. *Anal. Chim. Acta* **2010**, *663*, 60–68.
25. Keusgen, M.; Curtis, J. M.; Thibault, P.; Walter, J. A.; Windust, A.; Ayer, S. W. *Lipids* **1997**, *32*, 1101–1012.

## Chapter 8

# Carotenoid Content And Flavor Of Pumpkin Juice

**H. Dietrich,* M. Hey, C. Patz, P. Kürbel, and B. Froehling**

**Department of Wine Analysis and Beverage Technology, Geisenheim Research Center, von-Lade-Strasse 1, D-65366 Geisenheim, Germany**
***E-mail: Helmut.Dietrich@fa-gm.de.**

Pumpkin juice and pomace are rich sources of carotenoids. Their composition and concentration is strongly dependent on the pumpkin varieties and the kind of processing. It was found that a two-step-milling (classical mill + Supraton) is a prerequisite to reach an economic process in regard to juice with a concomitant increase of carotenoids. An additional extraction of pumpkin pomace can also be recommended as a useful measure for increasing valuable bioactive substances.

Significant differences were shown for the carotenoid content ranging from 8 mg/L to 173 mg/L. The soluble solids of the juices show a very broad range from 1.88 – 13.43 °Brix, corresponding to the total sugar content of 7.8 to 60.8 g/L. The product is also a rich source for minerals, especially potassium which ranges from 2246 to 4924 mg/L. Due to low sugar content, pumpkin juice could be an appropriate source for low-calorie-beverages. Several combinations with fruit juices and purees were tested leading to products with a good sensory evaluation.

## Introduction

The genus *Cucurbita* belongs to the family *Cucurbitaceae* and comprises ca. 35 species. Thereof, 5 species comprising many varieties are the most important: *Cucurbita maxima* (Giant pumpkins); *Cucurbita pepo* (vegetable pumpkin, Zucchini); *Cucurbita moschata* (Muscat pumpkin); *Cucurbita ficifolia*, and *Cucurbita mixta* (Ayote).

Pumpkin (*C. maxima*) is used as decoration (Halloween), as vegetable, for creamy soups, as pumpkin seed preparations against prostate and bladder complaints, for extraction of edible oil from seeds [unsaturated fatty acids ~70%, phytosterols] and as raw material for food industry (coloring matter, powders). Not much is known about the use of pumpkin juice.

The red and orange colored pumpkins *Cucurbita maxima L.* are valuable sources of carotenoids among vegetables or vegetable juices, especially those varieties with orange pulp. Murkovic et al. (*1*) focused their research on the quantification of α-carotene, β-carotene and lutein on different varieties of the species *C. pepo L.*, *C. maxima L.*, and *C. moschata L.*. The content of the carotenoids ranged from 0.6 to 74 mg/kg for β-carotene, from 0 to 75 mg/kg for α-carotene and from 0 to 170 mg/kg for lutein. Hidaka et al. (*2*) and Arima and Rodriguez-Amaya (*3*) found substantial differences in the carotenoid profile of several *Cucurbita* species as well. Arima und Rodriguez-Amaya (1988) found an average of 16.6 mg β-carotene/kg fresh weight, 10.2 mg/kg lutein, 8.2 mg/kg violaxanthin, 8.8 mg neoxanthin,2.4 mg/kg zeaxanthin and low amounts of α-carotene, ζ-carotene, cryptoxanthin, taraxanthin and luteoxanthin.

Pumpkin juices could be a valuable source of fruit-vegetable based beverages as they are characterized by low energy content (low sugar) and concomitant complex carotenoid profile. The restrictions are the taste and the tendency to form off-flavors.

The interest in new beverages with high fruit content and low energy is easily explained by the fact that most of the German consumers, especially children and young adults, do not eat enough fruits and vegetables (*4*, *5*). Moreover the daily sugar intake of children is a matter of concern worldwide. The key questions in this regard are: (i) is it possible to produce convenience products with high content of fruits&vegetables, and (ii) is it possible to hide the typical vegetable (pumpkin) aroma?

The aim of this study was the chemical characterization of pumpkin varieties as a basis for juice production and the development of fruit juice-pumpkin juice combinations and the evaluation of their sensory acceptance by young consumers (10-16 years). Also the processing should be improved for high yields and for a better preservation of carotenoids. In selected varieties, the distribution of carotenoids between fruit flesh and peel was determined.

## Materials and Methods

### Chemical Analysis

Chemical analysis as well as the analysis of the carotenoid profiles of juices, peel, and fruit flesh were performed as described elsewhere (*6*, *7*).

### Juice Processing

In the years 2003-2008 13 pumpkin varieties were processed in lots of about 200 kg each to optimize the processing line (*8*) and to select suitable varieties for beverage production. Based on these results the following procedure was used. Genuine pumpkin juices were produced on a half technical scale (lots of 150 kg) of each of the following three pumpkin varieties *Rouge*, *Neon*, *Muscat*, due to good sensory ratings. The processing consisted of washing, manual pre-disintegration, manual seed removal, one step grinding with a seepex-pump®, in line homogenisation (SUPRATON®) for a fine dispersion of the pumpkin mash, mash heating for enzyme deactivating over a spiral flow (90 °C), cooling over a spiral flow to 50 °C and immediate enzyme treatment with 100 mL/t cellulase and pectinase (Erbslöh, Geisenheim) for 60 min. Subsequently juice and pomace were separated with a decanter (Flottweg Z23). After adding 3 g/L citric acid and 200 mg ascorbic acid, juices were bottled and pasteurised. Besides juice production (A-juice) a pomace extraction was made. Therefore, water (2:1) and enzymes (100 mL/t pectinase and cellulase, Fa. Erbslöh, Geisenheim) were added at 50 °C for 60 min, followed by decanter separation. 3 g/L citric acid and 200 mg/L ascorbic acid were added, the pomace extracts were filled and pasteurised in bottles.

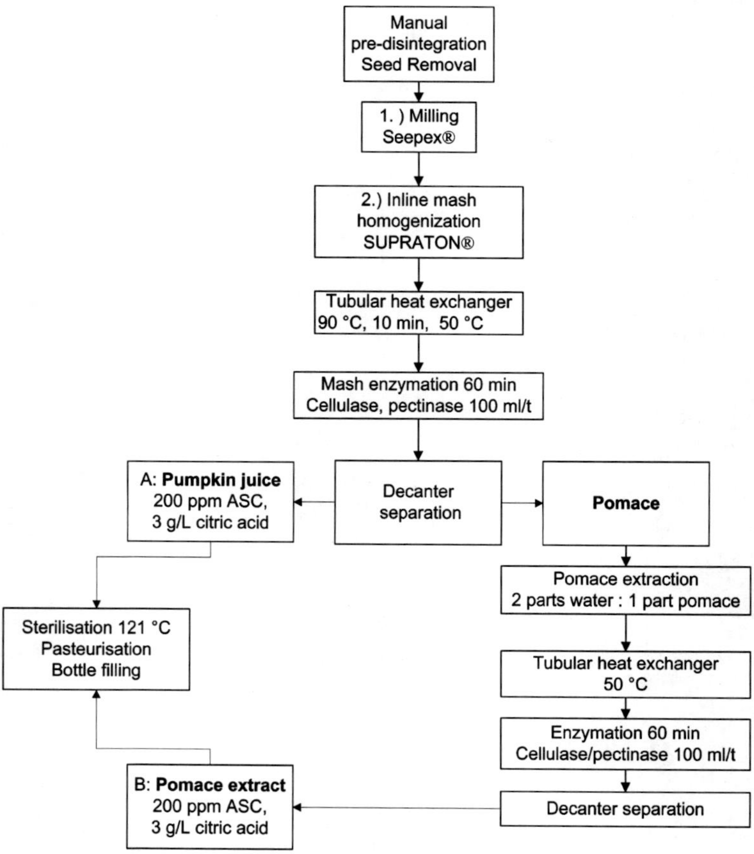


*Figure 1. Processing of pumpkins into pumpkin juice (A) and pomace extract (B).*

## Beverage Development Based on Fruits and Vegetables

For pumpkin juice preparation, which turned out to be a major challenge for beverage technologists, two strategies were followed:

1. Development of brilliant-yellow juices by combining pumpkin juice with other yellowish fruit juices
2. Abandon the yellow color and "hide" pumpkin juice behind intensely red juices

The combination of pumpkin juice with red fruit juices would result in beverages with high polyphenol/anthocyanin content and high carotenoid content.

# Results and Discussion

## Influence of Processing on Juice Yield and Carotenoid Content

Within several years, the processing of cloudy pumpkin juices was ameliorated with the aim to increase juice yield and carotenoid content. As is the case also for carrot juice it was found that a two-step-milling is a prerequisite to reach an economic process. An example is shown in Figure 2. Milling with a seepex mill or raetz mill results in yields between 68.6% and 78.9%, resp. with carotenoid contents up to 60.7 mg/L. The additional use of a second step (Supraton rotor-stator-system) increases the juice yield to 83.2% with a concomitant increase of carotenoids to 109.5 mg/L, which means nearly a doubling of these bioactive compounds.

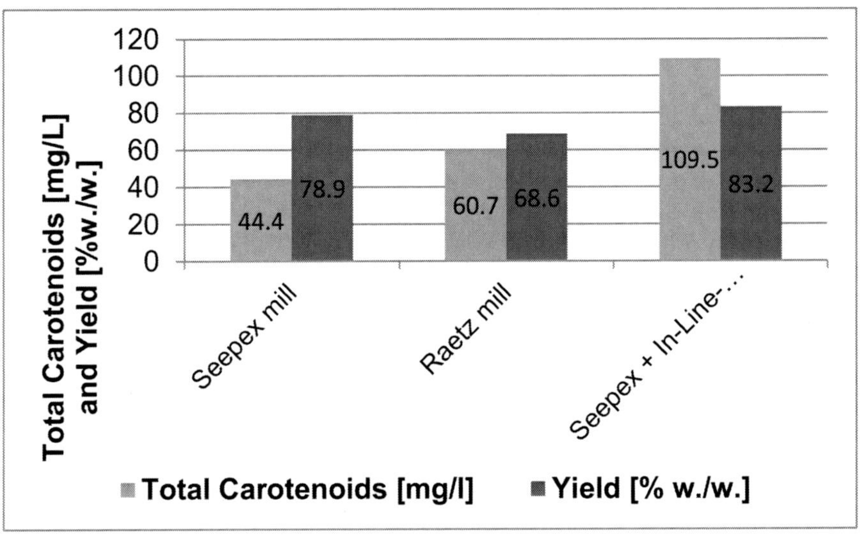


*Figure 2. Influence of cell disrupture with one-step milling (Seepex or Raetz mill) in comparison to a two-step cell disrupture (Seepex + in-line homogenization SUPRATON®).*

**Table 1. Composition of Pumpkin Juice (A-juice) and Pomace Extract (B-juice), Acidified with Citric Acid and Related to 8 °Brix**

| | *Soluble solids* | *Total sugar* | *Sugar free extract* | *pH* | *Total acidity* | *Ash* | *K* | *Ca* | *Mg* | *Polyphenols* |
|---|---|---|---|---|---|---|---|---|---|---|
| | *° Brix* | *g/l* | *g/l* | | *g/l* | *g/l* | *mg/l* | *mg/l* | *mg/l* | *mg/l* |
| *Muscat* A | 7.22 | 50.4 | 39.6 | 4.41 | 4.29 | 7.2 | 4080 | 173 | 144 | 286 |
| *Muscat* B | 3.07 | 40.7 | 47.7 | 3.79 | 3.31 | 5.9 | 2676 | 185 | 133 | 406 |
| *Rouge* A | 5.05 | 37.9 | 54.3 | 3.97 | 4.36 | 10.5 | 6528 | 464 | 201 | 504 |
| *Rouge* B | 2.47 | 37.6 | 54.4 | 4.33 | 3.5 | 10.6 | 4644 | 460 | 198 | 554 |
| *Neon* A | 4.68 | 35.6 | 57.8 | 4.31 | 4.26 | 11.8 | 6761 | 301 | 318 | 553 |
| *Neon* B | 1.94 | 32.2 | 60.2 | 3.77 | 3.54 | 10.8 | 4705 | 293 | 313 | 696 |

In our early studies, the seeds were not removed. Later, we found out that seeds should be removed as much as possible to avoid aroma off-flavors (see Figure 1).

An additional extraction of pumpkin pomace, as shown in Figure 1, is a useful measure for increasing valuable substances, like sugars, minerals and polyphenols. This is shown in Table 1 for the varieties *Muscat*, *Rouge* and *Neon*. The high acidity is explained by the previous addition of citric acid.

It can be concluded that processing has a strong impact on economy, chemical composition and quality of pumpkin juices.

## Chemical Characterization of Pumpkin Juice

The chemical composition of pumpkin juices is shown in Table 2 for pumpkin juices which were not yet acidified with citric acid. The soluble solids show a very broad range from 1.88 – 13.43° Brix, corresponding to the total sugar content of 7.8 to 60.8 g/L. Figure 3 shows that the sugars consist of glucose, fructose and saccharose. In comparison to fruit juices the sugar content is much lower. Therefore, pumpkin juice could be an appropriate source for low-calorie-beverages. The acidity, mainly due to malic acid, is very low with a concomitant high pH. As in the case of other vegetable juices, also in pumpkin juice production it is important to ensure microbial stability. The product is a rich source for minerals, especially potassium which ranges from 2246 to 4924 mg/L (mean 3861 mg/L). Calcium and magnesium are also present in significant amounts, whereas sodium is very low. The total polyphenols (Folin) range from 180 to 839 mg/L (mean 497 mg/L).

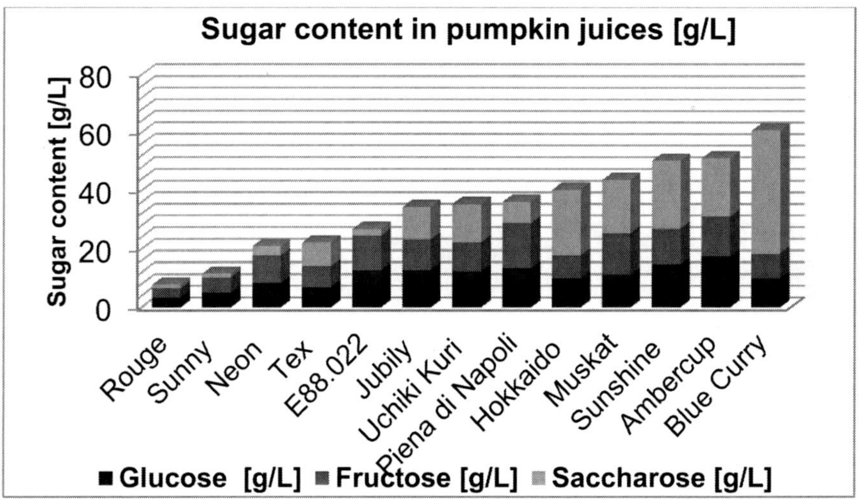


*Figure 3. Sugar composition of monovarietal pumpkin juices (n=13). Reproduced with permission from reference (8). Copyright (2011) (Confructa-Medien).*

**Table 2. Chemical Composition of Monovarietal Pumpkin Juices (n=13)**

| *N= 13 varieties* | *Mean* | *Minimum* | *Maximum* |
|---|---|---|---|
| Relative density d20/20 | 1.03251 | 1.00880 | 1.05774 |
| Soluble solids °Brix | 7.60 | 1.88 | 13.43 |
| Extract [g/L] | 84.3 | 22.7 | 149.9 |
| sugar free extract [g/L] | 50.3 | 14.9 | 89.1 |
| Total sugar [g/L] | 34.0 | 7.8 | 60.8 |
| Glucose [g/L] | 10.7 | 3.5 | 17.6 |
| Fructose [g/L] | 10.0 | 3.2 | 15.5 |
| Sucrose [g/L] | 13.3 | 1.1 | 42.3 |
| pH - value [g/L] | 6.3 | 5.8 | 6.8 |
| Titratable acidity [g/L] | 0.4 | 0.3 | 0.6 |
| L-lactic acid [g/L] | 0.05 | 0.05 | 0.05 |
| L-malic [g/L] | 2.1 | 0.7 | 3.8 |
| volatile acid [g/L] | 0.07 | 0.03 | 0.10 |
| Copper [mg/L] | 0.60 | 0.10 | 1.10 |
| Iron [mg/L] | 1.47 | 0.50 | 2.40 |
| Zinc [mg/L] | 0.75 | 0.09 | 1.20 |
| Sodium [mg/L] | 3.7 | 1.0 | 8.0 |
| Calcium [mg/L] | 162 | 129 | 209 |
| Potassium [mg/L] | 3861 | 2246 | 4924 |
| Magnesium [mg/L] | 189 | 44 | 300 |
| Nitrate [mg/L] | 312 | 50 | 584 |
| Chloride [mg/L] | 1056 | 434 | 1554 |
| Phosphate [mg/L] | 1049 | 386 | 1539 |
| Total polyphenols mg/L] | 497 | 180 | 839 |

## Carotenoids

The carotenoids cover a broad concentration range as shown in Figure 4. In the juices of the varieties *Rouge* (8.6mg/L), *Neon* (8.4 mg/L) and *Sunny* (10.9 mg/L), the concentrations were low, whereas the juices of the varieties *Hokkaido* (111.6 mg/L), *Sunshine* (136.4 mg/L), *Uschiki Kuri* (143.6 mg/L), *Ambercup* (145 mg/L) and *Jubily* (173.2 mg/L) were rich in carotenoids. The last mentioned varieties are comparable with commercial carrot juices in regard to carotenoid content.

The carotenoid profile of the pumpkin juices is composed of carotenes and xanthophylls. In dependence of the variety different concentrations and distributions of α-carotene, β-carotene, violaxanthin, neoxanthin, all-trans-lutein, zeaxanthin, lutheoxanthin and the isomers 9-cis-β-carotene and 13-cis-β-carotene were obtained as described by Kreck et al. (*7*).

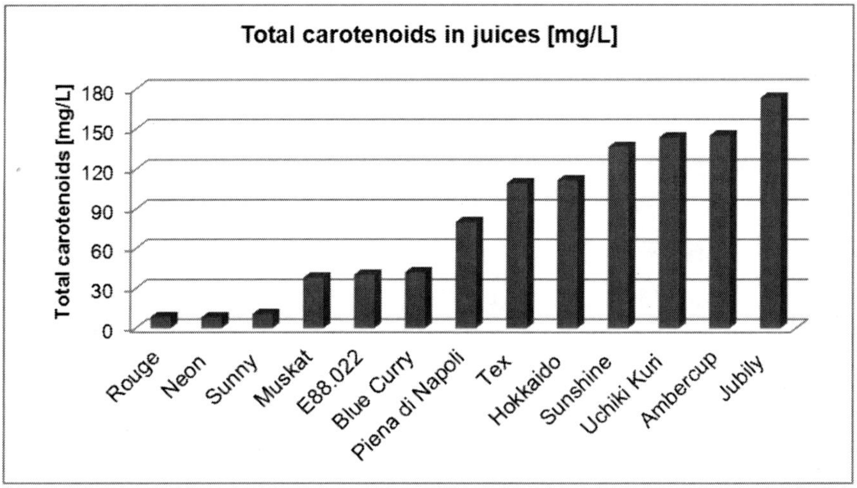


*Figure 4. Carotenoid concentration of pumpkin varieties (n=13). Reproduced with permission from reference (8). Copyright (2011) (Confructa-Medien).*

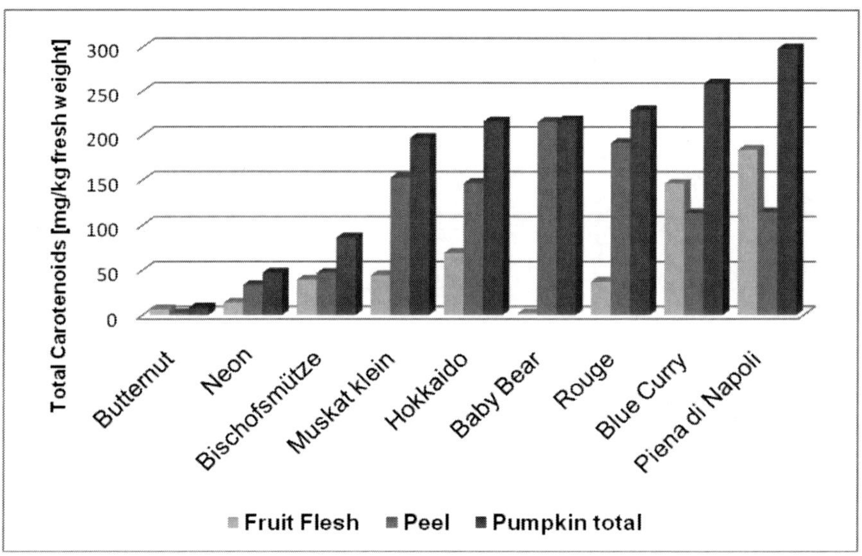


*Figure 5. Carotenoid distribution in fruit flesh and peel (n=9). Reproduced with permission from reference (8). Copyright (2011) (Confructa-Medien).*

The carotenoids are unequally distributed between peel and fruit flesh which could also be a reason for the differences in pumpkin juices. In 7 varieties, we found 5.9 – 10.5 % of the carotenoids in the peels, 67.9 – 85.7 % in the flesh and 6.6 – 22.7% in the pomace, on a fresh weight basis (*7*).

The distribution between peel and flesh of another series of varieties is shown in Figure 5.

## Aroma Profile

The results of the flavor analysis showed also distinct differences between the varieties as previously described (*6*). Whereas the juice of the variety *Rouge* was dominated by C9-compounds (nonanal, nonen-1-ol) and the norisoprenoid β-damascenon, the main compounds of *Muscat* were identified as β-ionone, α-ionone, dihydropseudoionone and β-cyclocitral. The juice of the variety *Sunny* was dominated by C6-compounds (2-hexenal, hexanal), hexanoic acid ethyl ester and 1-octen-3-ol. From a sensory point of view the *Muscat* variety was rated best.

Not every pumpkin variety would be appropriate for the beverage development, as they contain typical cucumber-like aroma C9- compounds, probably due to lipoxygenase-catalysed reactions of unsaturated fatty acids. This was even the case when the seeds were removed. For beverage development it would be necessary in some cases to remove the aroma and to use dearomatized pumpkin juice concentrate.

For these reasons, pumpkin juices possess a very versatile aroma. Figure 6 shows exemplarily the descriptive sensory evaluation of four varieties which are well suited for beverages due to sensory properties and carotenoid content.

## Beverage Development Based on Fruits and Vegetables

Beverage development is directed to products with low sugar content, brilliant and stable color, a pleasant taste and aroma, and high concentration of secondary plant metabolites, like carotenoids and polyphenols. Pumpkin juices could be a good source but on the other hand, they possess a complex aroma which tends to fast aging and off-flavor formation. Two strategies which are described here briefly for combining with other juices were pursued:

### *1. Strategy: Development of Yellow Fruit Juice Blends*

Several compositions were tested based on fruit juices without anthocyanins to get juice blends with an intense yellow color. Details are described in ref. (*9*).

*Multicomponent Juices*

These products were based on different juice compositions of apple, orange, sea buckthorn, banana, mango puree, maracuja, pineapple and peach with the addition of with dearomatized pumpkin juice concentrate (30%) or pumpkin direct juice cv. *Muscat* 15%.

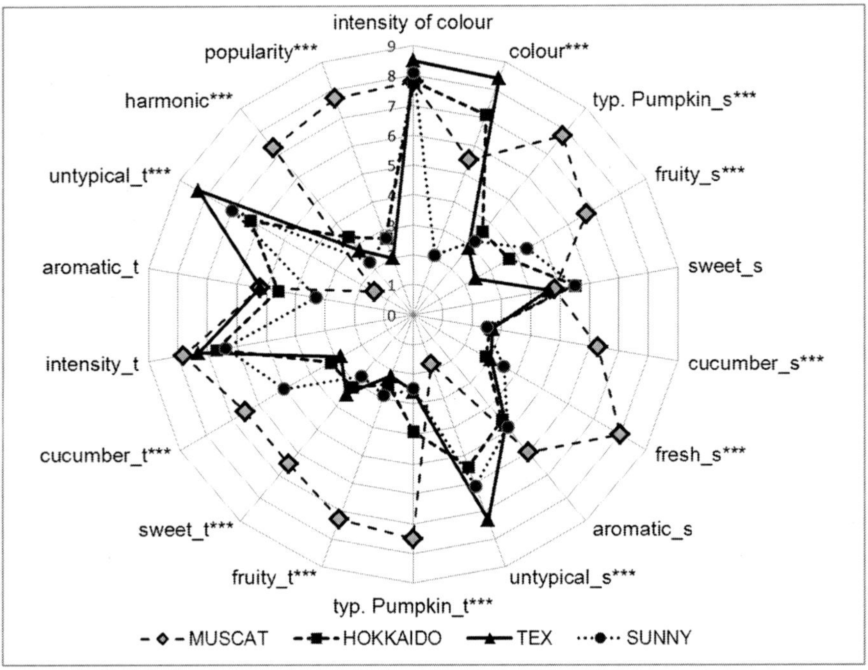


*Figure 6. Sensory description of pumpkin juices made from the varieties Muscat, Hokkaido, Tex and Sunny. t: taste, s smell, *** significance 99%.*

It turned out that these products were too complex without a clear taste and aroma profile which were not regarded as positive by the sensory panel. The products showed a fast aging and were sensitive against oxidation which limits the shelf life.

Better results were obtained using apple and orange juices.

*Apple-Pumpkin-Orange Juice*

1.1 cloudy apple juice 50%; dearomatized pumpkin juice from concentrate (30%), orange juice from concentrate 20%

1.2 cloudy apple juice 50%; pumpkin direct juice cv. *Muscat* 25% (including the native aroma), orange juice concentrate 25%.

The sensory evaluation revealed that apple-pumpkin-orange juice with pumpkin direct juice cv. "Muscat" 25% was the favorite.

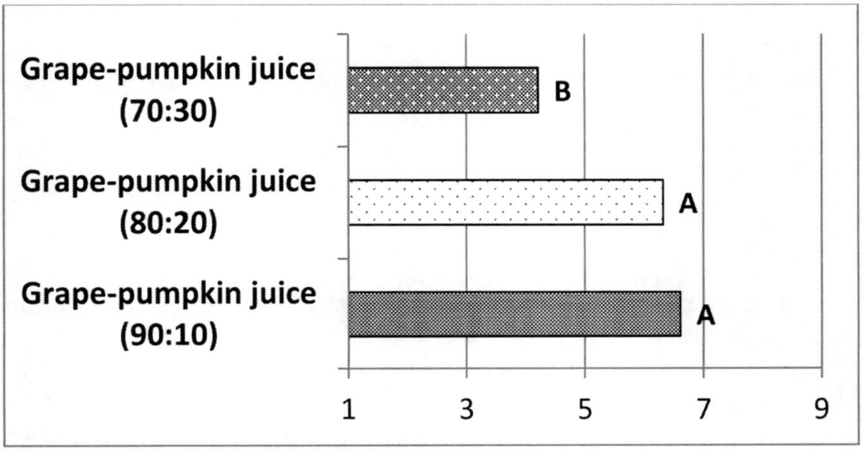


*Figure 7. Acceptance test of grape juice and pumpkin juice blends on a scale from 1-9. ANOVA.*

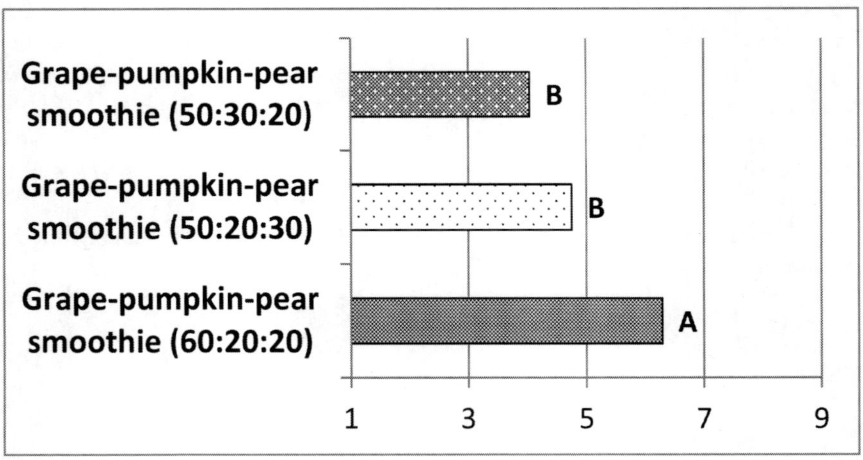


*Figure 8. Acceptance test of grape-pumpkin blends with pear puree on a scale from 1-9. ANOVA.*

*2. Strategy: Masking Behind Anthocyanin-Rich Grape Juices*

Pumpkin juice can also be combined with red colored juices with a high concentration of anthocyanins. In these blends the yellow color of carotenoids is no more visible.

Grape juice cv. *Dakapo* was blended with 10%, 20% and 30% pumpkin juice (*Hokkaido*), reconstituted from concentrate to 4.5°Brix. The products were evaluated by a sensory panel of 28 children between 10 and 16 years. The result of the acceptance test in Figure 7 shows that the products with 10% and 20% were assessed likewise, but the product with 30% was evaluated significantly worse ($p > 0.05$).

In order to increase the dietary fiber content, the grape-pumpkin juice blend was added with pear puree (20-30%). Figure 8 shows that the blend 60:20:20 of grape-pumpkin-pear puree was significantly better than the two other products.

## Conclusions

Pumpkin juices have low sugar contents, low acid, high mineral contents and are a rich source of carotenoids depending on the variety. The carotenoid composition is complex and unequally distributed between flesh (pulp) and peel. In dependence of the variety different concentrations and distributions of α-carotene, β-carotene, violaxanthin, neoxanthin, all-trans-lutein, zeaxanthin, lutheoxanthin and the isomers 9-cis-β-carotene and 13-cis-β-carotene were obtained.

The flavor (off-flavor) is the major problem in beverage development. The aroma is very complex depending on the variety. Main components are C9 / C6-compounds and C13-norisoprenoids. Varieties with a high carotenoid content are not necessarily the best choice for beverage development, due to sensory restrictions. This was one of the reasons to select *Rouge*, *Neon* and *Muscat* pumpkin juices in the beginning of the study, in spite of low carotenoid contents. A way to circumvent this problem is the use of dearomatized concentrate. This would allow the application of other varieties with high carotenoid contents.

It is possible to combine a certain amount of pumpkin juice (mostly 10-20 %, in some cases 30 %) with aromatic fruit juices leading to products with significant proportions of carotenoids and anthocyanins / polyphenols. These products are low in calories and have a good sensory acceptance.

## References


1. Murkovic, M.; Muelleder, U.; Neunteufl, H. *J. Food Compos. Anal.* **2002**, *15*, 633–638.
2. Hidaka, T.; Anno, T.; Nakatsu, S. *J. Food Biochem.* **1987**, *11*, 59–68.
3. Arima, H. K.; Rodriguez-Amaya, D. B. *J. Micronutr. Anal.* **1988**, *4*, 177–191.
4. Yngve, A.; Wolf, A.; Poortvliet, E.; Elmadfa, I.; Brug, J.; Ehrenblad, B.; Franchini, B.; Haraldsdóttir, J.; Krølner, R.; Maes, L.; Pérez-Rodrigo, C.; Sjöström, M.; Thórsdóttir, I.; Klepp, K.-I. *Ann. Nutr. Metab.* **2005**, *49*, 236–245.
5. Diethelm, K.; Jankovic, N.; Moreno, L. A.; Huybrechts, I.; De Henauw, S.; De Vriendt, T.; González-Gross, M.; Leclercq, C.; Gottrand, F.; Gilbert, C.

C.; Dallongeville, J.; Cuenca-Garcia, M.; Manios, Y.; Kafatos, A.; Plada, M.; Kersting, M. *Public Health Nutr.* **2012**, *15*, 386–398.

6. Kreck, M.; Patz, C. D.; Ludwig, M.; Degenhardt, A.; Paschold, P.; Dietrich, H. *Dtsch. Lebensm-Rundsch.* **2004**, *100*, 445–452.
7. Kreck, M.; Kürbel, P.; Ludwig, M.; Paschold, P.; Dietrich, H. *J. Appl. Bot.* **2006**, *80*, 93–99.
8. Hey, M.; Patz, C. D.; Ludwig, M.; Dietrich, H. *Flüss. Obst.* **2011**, *78*, 60–65.
9. Muller A. Bachelor Thesis, University of Applied Science, Wiesbaden, Germany, 2007.

# Carotenoid Cleavage Products in Grapes and Wines

## Chapter 9

# Biosynthesis of $C_{13}$-Norisoprenoids in *Vitis vinifera*: Evidence of Carotenoid Cleavage Dioxygenase (CCD) and Secondary Transformation of Norisoprenoid Compounds

**Ziya Günata***

**Université Montpellier 2, UMR Qualisud, place E. Bataillon, 34095 Montpellier Cedex 5, France**
***E-mail: zgunata@univ-montp2.fr.**

Although the chemistry and occurrence of volatile norisoprenoids in plants is well studied little is known about their biosynthesis. This chapter gives an overview on the biosynthesis of $C_{13}$-norisoprenoids in grape berries, main class of norisoprenoids in *Vitis vinifera*. Experiments supportive for an apo-carotenoid pathway, the evidence of a carotenoid cleavage dioxygenase (CCD) from *Vitis vinifera* giving rise to $C_{13}$-norisoprenoids and the secondary transformations of $C_{13}$-norisoprenoids by grape cell culture are discussed.

Several $C_{13}$-norisoprenoids have been detected in grape berries and leaves (*1–4*). Some of them, such as β-damascenone, β-ionone, 1,1,6-trimethyl-1,2-dihydronaphthalene (TDN), Riesling acetal, vitispirane, and actinidol are potent aroma contributors in both red and white wines (*2*, *5–7*). In contrast to terpenes, $C_{13}$-norisoprenoids occur mainly as glycoconjugates in grape berry and wine leaves (*4*, *8*). This chapter discusses the biogenesis of $C_{13}$-norisoprenoids in grape berries and is presented in the following order: (i) arguments in favor of an apo-carotenoid pathway, (ii) evidence of a carotenoid cleavage dioxygenase (CCD), and (iii) secondary transformations of $C_{13}$-norisoprenoids.

## Arguments in Favor of an Apo-Carotenoid Pathway

Before the evidence of the existence of a CCD in *Vitis vinifera* (*9*) there were several assumptions in favor of an apo-carotenoid pathway to explain the origin of $C_{13}$-norisoprenoids in grapes as discussed below:

### Relation between the Levels of $C_{13}$-Norisoprenoids and Carotenoids during Grape Berry Maturation

During grape berry ripening from Muscat *cv.* the levels of $C_{13}$-norisoprenoids and carotenoids were found to change in opposite directions (*10*). A decline in carotenoid levels (β-carotene and xanthophylls) was observed from veraison to maturity, with the major reduction occurring at veraison. During the same period the levels of $C_{13}$-norisoprenoids increased with increasing maturity. In this study only the levels of $C_{13}$-norisoprenoids occurring as glycoconjugates were considered since their free counterparts were negligible in grape berries.

### Relation between Xanthophyll Cycle and $C_{13}$-Norisoprenoid Synthesis

Interconversions between different carotenoids taking part in the xanthophyll cycle under light and shaded conditions provided supplementary clues on the origin of norisoprenoids (*11*). Under light conditions, higher levels of 3-oxo-α-ionol, 3-hydroxy-7,8-dihydro-β-ionone and 3-hydroxy-7,8-dihydro-β-ionol were observed. This increase may be attributed to the non-epoxy xanthophylls, like zeaxanthin, lutein since their synthesis is favored under light. On the contrary, epoxyxanthophylls synthesis is favored under shade conditions producing higher levels of violaxanthin and neoxanthin. Higher levels of 3-hydroxy-5,6-epoxy-β-ionol, 3-hydroxy-7,8-dehydro-β-ionol and 4,5-dihydro-vomifoliol were also observed suggesting structural relationships between the norisoprenoids and their parent carotenoids. The exposure of grape berries to sunlight during maturation increased the level of $C_{13}$-norisoprenoids and decreased that of carotenoids compared to shaded berries. This supports the argument for an apo-carotenoid pathway for the synthesis of $C_{13}$-norisoprenoids (*11–13*). Furthermore increases in the levels of glycoconjugated TDN and Riesling acetal in white Riesling grapes were also observed when sunlight exposure greater than 20% of full sun exposure was applied beginning at veraison (*14*).

### Relation between the Stereochemistry of $C_{13}$-Norisoprenoids and Carotenoids

The comparison of the asymmetric centers of $C_{13}$-norisoprenoids with those of the putative grape carotenoids supported the hypothesis that carotenoids are precursors of norisoprenoids. Indeed, formation of megastigmane-3,9-diol and 3-oxo-α-ionol from the ε-ring of lutein was suggested since the stereochemistry of carbons-3 and -6 was maintained (*15*). Similarly formation of 3-hydroxy-β-damascone and β-damascenone from neoxanthin was proposed because of the maintenance of the stereochemistry of carbon-3 (*15–17*).

### $^{13}C$-Labelling Experiment

$^{13}C$-labelling experiments were performed to introduce $^{13}CO_2$ into berries during berry ripening in order to examine a possible apo-carotenoid pathway through the analysis of carbon isotopic ratios in carotenoids and norisoprenoids (*15, 18*) (Figure 1).

$^{13}CO_2$ labeling of berries was performed before veraison since carotenoid synthesis in grape berries occurs mainly prior to veraison (*10*). Grape bunches from Syrah cv. were hermetically maintained in the bags and were fed with $^{13}CO_2$ from 8 a.m to 3 p.m each day from June 16 to July 26. The feeding started 16 days following berry set, lasted for 40 days and ended 6 days before veraison. At the end of each $^{13}CO_2$ feeding, the bags were flushed with air. To trap humidity and hence avoid fungal development on the grapes the water in the circuit was cold-trapped. Samplings were done at two moments, 6 days before veraison (August 01) and at maturity (September 19). Levels and isotopic ratios of $^{13}C/^{12}C$ (δ, ‰) of β-carotene and xanthophylls from labeled and control berries were determined. In parallel the same determination was performed for 3-oxo-α-ionol, one of the major glycoconjugated $C_{13}$-norisoprenoid in grape.

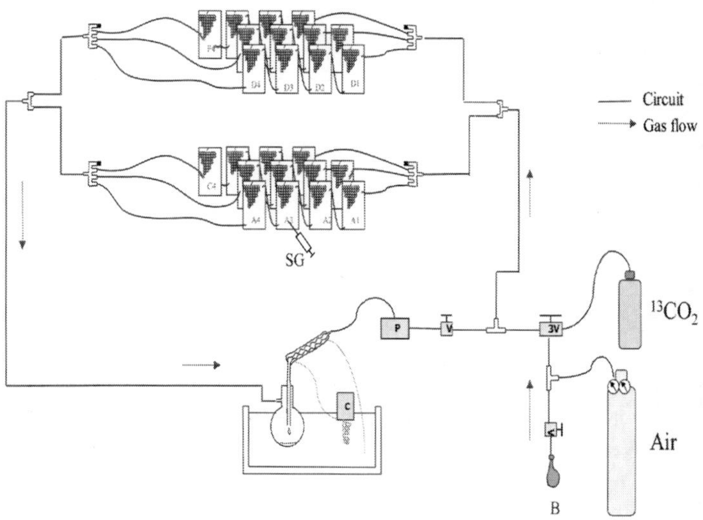


*Figure 1. $^{13}CO_2$ labeling experiment of grape berries : B, bag; C, cryostat (2-3°C); SG, gas syringe; V, valve; Ai-Fi (i=1-4), circuits (x6) of transparent bags (x4) containing grape bunches (one per bag) for six plants. (Reproduced with permission from reference (15). Copyright 2002 Elsevier.)*

$^{13}CO_2$ was incorporated into the carotenoids as shown in Table I. Labeling of β-carotene (δ, ‰ : +109) was greater than that of xanthophylls (δ, ‰ : +62). At maturity there was a decrease in $^{13}C$-labelling of both carotenoids that could be explained by some carotenoid synthesis from veraison to maturity since $^{13}CO_2$ feeding ended before veraison. However the data indicate that a major part of carotenoids was synthesized before veraison since δ ‰ values are quite high at maturity. Furthermore in agreement with the literature, the concentration of carotenoids decreased significantly from veraison to maturity (*10*).

$^{13}CO_2$ was also incorporated into 3-oxo-α-ionol. The isotopic ratio of 3-oxo-α-ionol clearly increased from veraison to maturity suggesting that this compound was generated from the labeled carotenoid precursor during maturation. At maturity δ ‰ value of this compound was close to that of β-carotene and xanthophylls. All data points to the carotenoidic origin of 3-oxo-α-ionol.

**Table I. Isotopic Ratios and Concentrations of Carotenoids and 3-Oxo-α-ionol in Control and $^{13}CO_2$ fed Syrah Berries. (Reproduced from reference (*15*))**

| *Berries* | *β-Carotene* | *Xanthophylls* | *3-Oxo-α-ionol* |
|---|---|---|---|
| δ-values (δ, ‰) | | | |
| Labeled and sampled 26 July | +109 | +62 | +13.1, +22.7 |
| Labeled and sampled 19 September | +71 | +55 | +58.6, +61.1 |
| Control 19 September | -32 | -32 | -31, -31.5 |
| Concentrations mg $kg^{-1}$ | | | |
| Labeled and sampled 26 July | 5.6 | 5.6 | 29 |
| Labeled and sampled 19 September | 1.9 | 1.9 | 74 |

## Evidence of a Carotenoid Cleavage Dioxygenase (VvCCD1) from *Vitis vinifera*

The first carotenoid cleavage oxygenase (CCO) giving rise to volatile norisoprenoids was only reported in 2001 (*19*). Through the heterologous expression in *Escherichia coli* of the relevant CCO gene from *Arabidopsis thaliana* the formation of $C_{13}$-norisoprenoids from several carotenoids was demonstrated. The cleavage was occurring at 9,10 (9′,10′) double bonds of carotenoids. This paper has stimulated studies in several plant systems for the existence of CCOs which give rise to norisoprenoids. Since then CCOs have been identified in several plants and fruits such as *Vitis vinifera* (*9*), *Crocus sativa* (saffron) (*20*), *Petunia hybrida* (*21*), *Lycopersicon esculentum* (tomato) (*22*), *Cucumis melo* (melon) (*23*), *Chyrysanthenum morifolium Ramat* (*24*), *Citrus cv.* (*25*), *Fragaria ananassa* L. (strawberry) (*26*), *Zea mays* (*27*), *Coffea arabica* and *Coffea canephore* (*28*), *Rosa damascena* (*29*) *Osmanthus fragrans* Lour (*30*) and *Prunus persica* (*31*). This class of CCO cleaves carotenoids at the 9,10 (9′,10′) double bonds often with a large substrate specificity. They are classified as carotenoid cleavage dioxygenase 1 (CCD1). A dioxygenase mechanism has been proposed for CCDs since two oxygen atoms from $O_2$ are incorporated into the cleavage products instead of one oxygen atom from $O_2$ and one from water (case of monooxygenases from mammals) (*31*). Some CCDs are prone to cleave 9,10 bonds of apocarotenoids and as well as 5,6 (5′,6′) and 7,8 (7′,8′) double bonds of carotenoid precursors but further studies are needed to confirm this (*32*).

With regard to *Vitis vinifera* L. one clone from EST collection showed a full-length cDNA (VvCCD1) exhibiting homologies with CCD1 from *Arabidopsis thaliana* (*AtCCD1*) (*9*). A cytoplasmic location of VvCCD1 was proposed since no chloroplastic transit peptide could be predicted (*9*). This is in agreement with the proposition of cytosolic localization of CCD1 in plants and fruits in contrast to all other CCDs (CCD4, CCD7 and CCD 8) and NCED (Nine-*cis*-epoxycarotenoid dioxygenase) localized in the plastids (*32*). Cloning of VvCCD1 into β-carotene, zeaxanthin or lycopene accumulating *E.coli* strain resulted only in the discoloration of the colonies from *E.coli* accumulating zeaxanthin (*33*) (Figure 2).

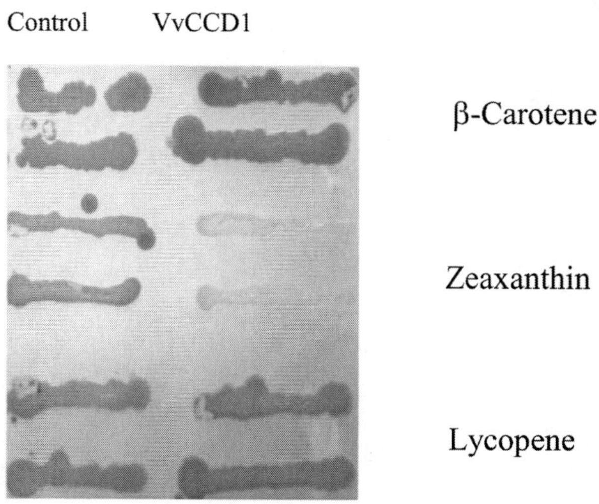


*Figure 2. Colonies from E.coli accumulating β-carotene, zeaxanthin, lycopene. VvCDD1 raw corresponds to the transformed E.coli with VvCCD1.*

A simple spectrophotometric enzyme assay was developed to measure CCD activity (*34*). Incubation of zeaxanthin with recombinant VvCCD1 from *E.coli* allowed to identify 3-hydroxy-β-ionone and $C_{14}$-dialdehyde by GC/MS. HPLC analysis indicated that VvCCD1 cleaves this carotenoid at the 9,10 (9′,10′) positions as shown in Figure 3.

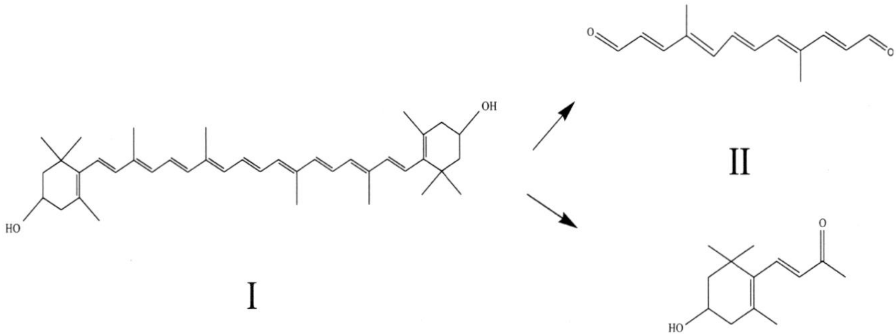


I

II

III

*Figure 3. Degradation products of zeaxanthin by VvCDD1. I: zeaxanthin; II: $C_{14}$-dialdehyde; III: 3-hydroxy-β-ionone.*

The same regiospecific cleavage by VvCCD1 occurred when lutein was used as substrate. Among the cleavage products 3-hydroxy-β-ionone and $C_{14}$-dialdehyde were detected (*34*). However neither β-carotene nor lycopene were accepted as substrates suggesting that carotenes are no substrate for this enzyme. The presence of β-ionone in grape and wine may then have several origins: (i) chemical oxidation of β-carotene (*35*), (ii) photooxygenation (*36*), (ii) co-oxidation mechanism under the action of lipoxygenase (*37*), and (iv) action of another grape CCD. In contrast to grape CCD1, recombinant CCD1 from *Arabidopsis thaliana* (*19*), from *Osmanthus fragrans* Lour. (*30*) were able to produce β-ionone from β-carotene. CCD1 from tomato was proposed to cleave β-carotene to yield β-ionone since silencing of the relevant CCD1 genes led to a significant decrease in the level of this flavor compound in ripe fruits (*22*). Similarly CCD1 from *Cucumis melo* was proposed to be involved in the formation of β-ionone from β-carotene through the studies involving up-regulation of the relevant gene (*23*).

## Expression of VvCCD1 During Grape Berry Development

Expression of VvCCD1 during grape berry ripening from two cultivars (Shiraz and Muscat) together with $C_{13}$-norisoprenoid levels have been studied (*9*). A significant induction of VvCCD1 expression (evidenced by real-time PCR) during the week prior to veraison was observed for both cultivars, with a two-fold induction for Muscat of Alexandria and a nearly four-fold one for Shiraz (Figure 4). Following veraison, the gene expression was almost stable during ripening. VvCCD1 expression level was higher in Muscat of Alexandria berries than in Shiraz. This could explain higher levels of $C_{13}$-norisoprenoids in the former. $C_{13}$-norisoprenoids were mainly detected in glyconjugated form and their level increased significantly after veraison. A strong increase was observed for the Muscat of Alexandria cultivar during the first week following veraison. Shiraz berries showed a progressive increase throughout ripening. Among eleven $C_{13}$-norisoprenoids detected, at maturity 4-oxo-β-ionol, 3-hydroxy-7,8-dihydro-β-ionone, 3-oxo-α-ionol, 3-hydroxy-β-ionone were the major ones. Only two primary cleavage products of carotenoids were detected : 3-hydroxy-β-ionone, a cleavage product of zeaxanthin or lutein, and 3-hydroxy-5,6-epoxy-β-ionone, a cleavage product of violaxanthin.

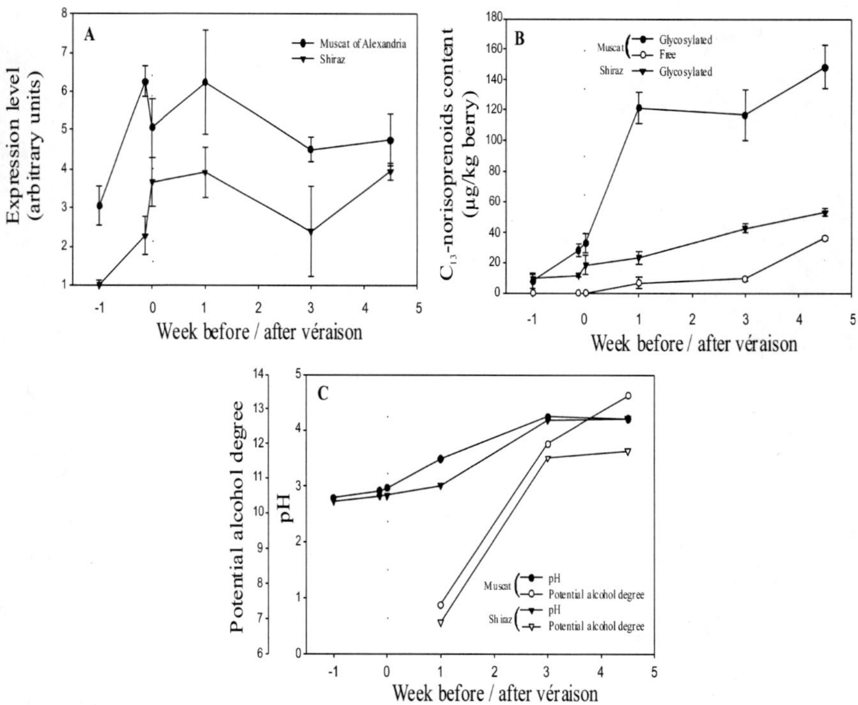


*Figure 4. Changes in the expression profile of VvCCD1 (A), in the levels of free and glycosylated $C_{13}$-norisoprenoids (B) and in pH and potential alcohol degree (C) in the berries from Muscat of Alexandria and Shiraz cultivars during grape berry development. Week 0 corresponded to veraison. (Reproduced with permission from reference (9). Copyright 2012 Oxford University Press.)*

## Secondary Transformations of $C_{13}$-Norisoprenoids

Several $C_{13}$-norisoprenoids with a large structural diversity have been identified in grape berries and wine leaves (1-4) suggesting that VvCCD1 cleavage products are prone to further transformations. A cell suspension culture of cv. Gamay was studied for its ability to transform administrated β-ionone and dehydrovomifoliol (*38*). Both norisoprenoids were metabolized, leading in each case to several norisoprenoidic compounds in free and glycoconjugated forms. Structures of metabolites detected in grape cell cultures administrated with β-ionone are shown in Figure 5. The same metabolites were identified in the grape cell cultures administrated with dehydrovomifoliol except for 3-oxo-α-ionone and 3-hydroxy-7,8-dihydro-β-ionone. Administered compounds were reduced at the side chain or oxygenated mainly at carbons 3 or 4 of the cyclohexane ring. The transformations suggest the involvement of hydroxylases, oxidoreductases and glycosyltransferases. Interestingly, most of the metabolites from the administrated compounds are present in grape berries and wine leaves (*3*, *4*).

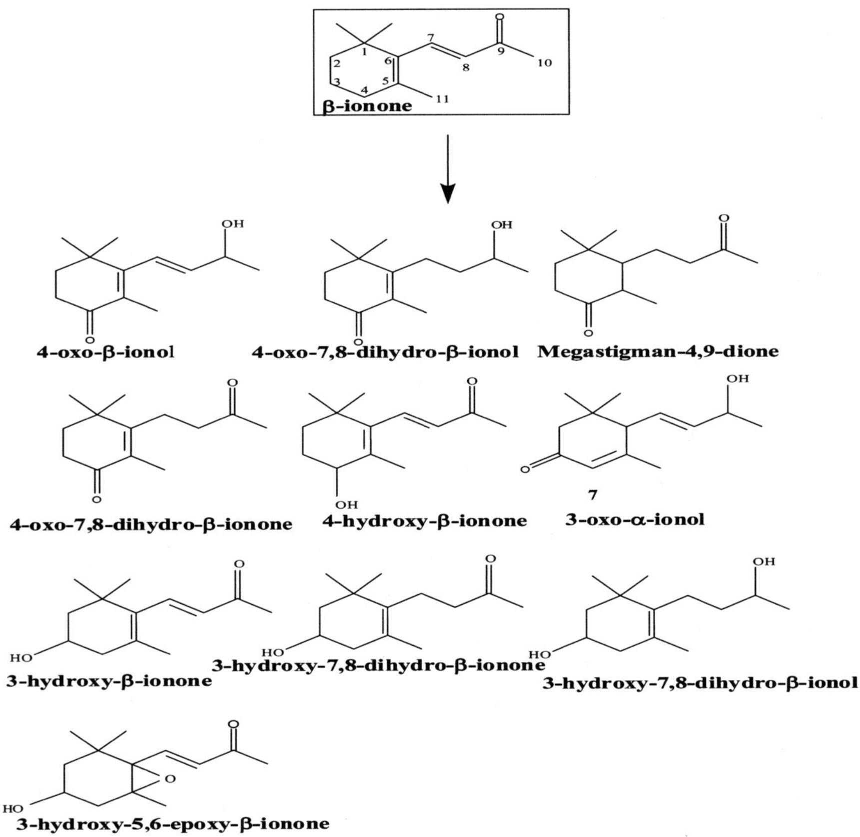


*Figure 5. $C_{13}$-norisoprenoids detected in grape cell cultures administrated with β-ionone. (Reproduced with permission from reference (38). Copyright 2012 Oxford University Press.)*

## Summary

A recombinant Carotenoid Cleavage Dioxygenase (CCD) from *Vitis vinifera* L. was able to symmetrically cleave zeaxanthin and lutein at the 9,10 (9′,10′) double bonds generating $C_{13}$-norisprenoids. On the contrary, β-carotene, amongst major carotenoids from grape was not a substrate. Expression of CCD gene was monitored by real-time PCR during berry development from Shiraz and Muscat of Alexandria cultivars. Gene expression was significantly induced approaching verasion. In parallel $C_{13}$-norisoprenoids level increased from verasion to maturity. $C_{13}$-norisoprenoids were prone to further modifications since a cell suspension culture from grape transformed two targeted $C_{13}$-norisoprenoids to several free and glycoconjugated compounds.

## Acknowledgments

The author of this chapter thanks his co-authors, who are identified in the references.

## References

1. Strauss, C. R.; Wilson, B.; Williams, P. J. *Phytochemistry* **1987**, *26*, 1995–1997.
2. Williams, P. J.; Sefton, M. A.; Francis, I. L. In *Flavor Precursors-Thermal and Enzymatic Conversions*; Teranishi, R., Takeoka, G., Günther, M., Eds.; ACS Symposium Series 490; American Chemical Society: Washington, DC, 1992; pp 74−86.
3. Skouroumounis, G. K.; Winterhalter, P. *J. Agric. Food Chem.* **1994**, *42*, 1068–1072.
4. Wirth, J.; Guo, W.; Baumes, R. L.; Günata, Z. *J. Agric. Food Chem.* **2001**, *49*, 2917–2923.
5. Winterhalter, P.; Rouseff, R. L. In *Carotenoid-derived Aroma Compounds*; Winterhalter, P., Rouseff, R. L., Eds.; ACS Symposium Series 802; American Chemical Society: Washington, DC, 2002; pp 1−17.
6. Skouroumounis, G. K.; Sefton, M. A. In *Carotenoid-derived Aroma Compounds*; Winterhalter, P., Rouseff, R. L., Eds.; ACS Symposium Series 802; American Chemical Society: Washington, DC, 2002; pp 241−254.
7. Marais, J.; Versini, G.; van Wyk, J.; Rapp, A. *S. Afr. J. Enol. Vitic.* **1992**, *13*, 71–77.
8. Gunata, Z.; Bayononve, C.; Baumes, R. L.; Cordonnier, R. E. *J. Chromatogr.* **1985**, *331*, 83–90.
9. Mathieu, S.; Terrier, N.; Procureur, J.; Bigey, F.; Gunata, Z. *J. Exp. Bot.* **2005**, *56*, 2721–2731.
10. Razungles, A.; Günata, Y. Z.; Pinatel, S.; Baumes, R.; Bayonove, C. *Sci. Aliments* **1993**, *13*, 59–72.
11. Razungles, A.; Baumes, R.; Dufour, C.; Sznaper, C.; Bayonove, C. *Sci. Aliments* **1998**, *18*, 361–373.

12. Bureau, S.; Razungles, A.; Baumes, R.; Bayonove, C. *Vitic. Enol. Sci.* **1998**, *53*, 64–71.
13. Bureau, S.; Baumes, R.; Razungles, A. *J. Agric. Food Chem.* **2000**, *48*, 1290–1297.
14. Gerdes, M: S.; Winterhalter, P.; Ebeler, S. E. In *Carotenoid-derived Aroma Compounds*; Winterhalter, P., Rouseff, R. L., Eds.; ACS Symposium Series 802; American Chemical Society: Washington, DC, 2002; pp 262−272.
15. Baumes, R.; Wirth, J.; Bureau, S.; Gunata, Y. Z.; Razungles, A. *Anal. Chim. Acta* **2002**, *458*, 3–14.
16. Skouroumounis, G. K.; Massy-Westropp, R.; Sefton, M. A.; Williams, P. J. *Tetrahedron Lett.* **1992**, *33*, 3533–3536.
17. Skouroumounis, G. K.; Winterhalter, P. *J. Agric. Food Chem.* **1994**, *42*, 1068–1072.
18. Wirth, J. Ph.D. Thesis, Montpellier University 2, Montpellier, Herault, France, 2001.
19. Schwartz, S. H.; Qin, X.; Zeevart, J. A. *J. Biol. Chem.* **2001**, *276*, 25208–25211.
20. Bouvier, F.; Suire, C.; Mutterer, J.; Camara, B. *Plant Cell* **2003**, *15*, 47–62.
21. Simkin, A. J.; Underwood, B. A.; Auldridge, M.; Loucas, H. M.; Shibuya, K.; Schmelz, E.; Clark, D. G.; Klee, H. J. *Plant Physiol.* **2004**, *136*, 3504–3514.
22. Simkin, A. J.; Schwartz, S. H.; Auldridge, M.; Taylor, M. G.; Klee, H. J. *Plant J.* **2004**, *40*, 882–892.
23. Ibdah, M.; Azulay, Y.; Portony, V.; Wasserman, B.; Bar, E.; Meir, A.; Burger, Y.; Hirschberg, J.; Schaffer, AA.; Katzir, N.; Tadmor, Y.; Lewinsohn, E. *Phytochemistry* **2006**, *67*, 1579–1589.
24. Ohmiya, A.; Kishimoto, S.; Aida, R.; Yoshioka, S.; Sumitomo, K. *Plant Pysiol.* **2006**, *3*, 1193–1201.
25. Kato, M.; Matsumoto, H.; Ikoma, Y.; Okoda, H.; Yano, M. *J. Exp. Bot.* **2006**, *57*, 2153–2164.
26. Garcia-Limones, C.; Schnabele, K.; Blanco-Portales, R.; Bellido, M. L.; Caballero, J. L.; Schwab, W.; Munoz-Blanco, J. *J.Agric. Food Chem.* **2008**, *56*, 9277–9285.
27. Vogel, J. T.; Tan, B-C.; MacCarthy, D. R.; Klee, H. J. *J. Biol. Chem.* **2008**, *283*, 11364–11373.
28. Simkin, A. J.; Moreau, H.; Kuntz, M.; Pagny, G.; Lin, C.; Tanksley, S.; McCarthy, J. *J. Plant Physiol.* **2008**, *165*, 1087–1106.
29. Huang, F. C.; Horvath, G.; Molnar, P.; Turcsi, E.; Deli, J.; Schrader, J.; Sandmann, G.; Schmidt, H.; Schwab, W. *Phytochemistry* **2009**, *70*, 457–464.
30. Baldermann, S.; Kato, M.; Kurosawa, M.; Kurobayashi, Y.; Fujita, A.; Fleischmann, P.; Watanabe, N. *J. Exp. Bot.* **2010**, *61*, 2967–2977.
31. Brandi, F.; Bar, E.; Mourgues, F.; Horvath, G.; Turcsi, E.; Giuliano, G.; Liverani, A.; Lewinsohn, E.; Rosati, C. *BMC Plant Biol.* **2011**, *11*, 11–24.
32. Walter, M. H.; Floss, D. S.; Strack, D. *Planta* **2010**, *232*, 1–17.
33. Mathieu, S. Ph.D. Thesis, Montpellier University 2, Montpellier, Herault, France, 2005.
34. Mathieu, S.; Bigey, F.; Procureur, J.; Terrier, N.; Gunata, Z. *Biotechnol. Lett.* **2005**, *29*, 837–841.

35. El Tinay, A. H.; Chichester, C. O. *J. Org. Chem.* **1970**, *35*, 2290–2293.
36. Isoe, S.; Hyeon, SB.; Sakan, T. *Tetrahedron Lett.* **1969**, 279–281.
37. Grosch, W.; Laskawy, G.; Kaiser, K. P. *Z. Lebensm.-Unters. Forsch.* **1977**, *165*, 77–81.
38. Mathieu, S.; Wirth, J.; Sauvage, F-X.; Lepoutre, J-P.; Baumes, R.; Gunata, Z. *Plant Cell, Tissue, Organ Cult.* **2009**, *97*, 203–213.

## Chapter 10

# Behavior of Glycosylated Monoterpenes, $C_{13}$-Norisoprenoids, and Benzenoids in *Vitis vinifera* cv. Riesling during Ripening and Following Hedging

**Imelda Ryona and Gavin L. Sacks***

**Department of Food Science, Cornell University, 630 West North Street, Geneva, New York 14456**
***E-mail: gls9@cornell.edu.**

Many key odorants in wines cannot be detected in winegrapes (*V. vinifera*) because they are present as glycosides. Although several studies on grape glycosides have been reported, the correlation of different glycoside classes with each other during ripening is not clear. Additionally, the effects of late season hedging on grape glycoside accumulation have not been studied. We characterized three major glycoside classes in Riesling grapes grown in New York State during the 2009 and 2010 seasons throughout ripening. Most glycosylated monoterpenes and $C_{13}$-norisoprenoids increased post-veraison, but the onset of accumulation was one to two weeks earlier for most $C_{13}$-norisoprenoids (>50% of maximum prior to 23 days post-veraison, 'dpv') than for monoterpenes (after 30 dpv). Within each aroma class, most monoterpene and $C_{13}$-norisoprenoid aglycones were highly correlated ($r>0.9$). The behavior of benzenoids was not consistent during the two seasons. Finally, late-season hedging of Riesling vines did not significantly affect the concentrations of most glycosylated volatiles at harvest.

# Introduction

Many wine volatiles are not found in a free state in winegrapes in appreciable quantity, but they are instead formed from non-volatile precursors ("bound volatiles") during fermentation and/or storage (*1*). Of these grape-derived precursors, the best-studied and most diverse class is arguably the β-glycosides, in which a volatile aglycone is bound to a sugar, e.g., a β-D-glucopyranoside (*1*). In mature wine grapes, glycosylated volatiles are generally at higher concentrations than their corresponding 'free' forms at harvest (*2*, *3*). During winemaking, the aglycone may be released either enzymatically by the action of yeast or lactic acid bacteria glycosidases (*4*) or non-enzymatically by acid hydrolysis (*5*, *6*), after which the released aglycone may undergo further transformations. While several volatile classes important to the aroma of finished wines are known to exist as glycosides in grapes, the two best studied grape glycoside classes are the monoterpenes, e.g., linalool, geraniol, and nerol, which contribute 'floral, Muscat' aromas; and the $C_{13}$-norisoprenoids, e.g., β-ionone, β-damascenone, which contribute a more diverse range of aromas. Beyond these two well-studied classes, lactones, alcohols, and volatile phenols, and in sum over 200 aglycones have been identified in grape glycoside extracts (*1*). Isolation of the glycoside fraction from grapes or wines is most commonly achieved by solid-phase extraction (SPE). The resulting glycosides can be analyzed intact by LC-MS or by GC-MS following derivatization, but they are more commonly subjected to either acid or enzymatic hydrolysis and the resulting aglycones isolated and analyzed by GC-MS (*7*). The aglycones released from acid hydrolysis are reportedly more similar to those released by fermentation than are the aglycones released by enzymatic treatment (*8*).

### Behavior of Glycosides during Grape Berry Ripening

Because of the importance of glycoside precursors to the aroma of finished wines, several groups have studied the evolution of different glycoside classes during grape berry maturation. Both monoterpene (*3*, *9*, *10*) and $C_{13}$-norisoprenoid (*11*) glycosides are reported to increase post-veraison, although the specific behavior can reportedly vary among cultivars or specific compounds. For example, the total $C_{13}$-norisoprenoid glycoside concentration is reported to increase dramatically one week post-veraison in Muscat before reaching a plateau but to increase more steadily between veraison and 4.5 weeks post-veraison in Syrah (*11*). The evolution of a limited number of glycosylated benzene derivatives in non-smoke tainted wines have been investigated (*10*), but their behavior has been inconsistent.

Previous studies concerning the evolution of volatile glycosides have generally focused on a single glycoside class (monoterpenes, $C_{13}$-norisoprenoids, and benzenoids). Since these reports are from different viticultural regions and often employ different cultivars, it is challenging to understand the relative timing of the evolution for the different glycoside classes. This information would be particularly useful for understanding the effects of harvest timing on cultivars like Riesling for which multiple glycoside classes, namely monoterpenes and

$C_{13}$-norisoprenoids, are important to the aroma of the varietal wine. Recent studies have profiled the evolution of free aroma compounds in Riesling and Cabernet Sauvignon during ripening (*12*, *13*), but a comparable profiling study of glycosylated volatiles has not been reported.

### Other Factors Affecting Glycoside Accumulation – Potential Role of Plant Wounding

Beyond grape maturity, viticultural factors that affect glycoside accumulation are still not well understood. It has been observed that grapes exposed to guaiacol and related volatile phenols from brushfires will accumulate these compounds as glycosides (*14*). Grape vegetative tissue will release a variety of volatile compounds when damaged, including monoterpenes and sesquiterpenes (*15*). These compounds may provide direct toxicity against herbivores, provide a signal to other plant tissues to increase the production of defense compounds, or recruit other insects to attack the herbivore, i.e., tritrophic interactions (*16*). Grapevines are physically wounded during hedging, in which the tops of the shoot tips are trimmed to obtain a desirable canopy architecture. Chemical elicitors which simulate insect feeding and plant wounding have been previously shown to induce production of other secondary metabolites relevant to flavor, such as some phenolics (*17*) and S-cysteine conjugates (*18*). Potentially, shoot tip hedging could increase grape berry glycoside concentrations by releasing volatiles which could then be absorbed and glycosylated, as is observed in 'smoke taint' or by inducing production of other volatiles in the grape berries.

### Objectives

Previous work has considered the evolution of different grape glycoside classes in isolation, but it has not considered the relative timing of these changes. The role of plant wounding has also not been considered. We profiled the evolution of multiple aglycone classes in *V. vinifera* cv. Riesling during ripening over two growing seasons using gas-chromatography time of flight mass spectrometry (GC-TOF-MS). The contribution of non-glycoside artifacts to the glycoside extract was also characterized. We also evaluated whether late season hedging resulted in changes in glycosylated precursors.

## Materials and Methods

### Chemicals

All solvents and aroma standards were purchased from Fisher Scientific (Pittsburgh, PA) and Sigma Aldrich (Allentown, PA), respectively, except for 1,1,6-trimethyl-dihydronaphthalene (TDN) which was synthesized in house using a protocol described elsewhere (*19*). Solid phase extraction (SPE) cartridges packed with 200 mg LiChrolut EN sorbent (Merck, Darmstadt, Germany) were purchased from VWR International (West Chester, PA).

## Vineyard Specifications

In 2007, Riesling (clone 110) on rootstock 3309 was planted with a spacing of 2.75 m between rows and 1.10 m between vines at a Cornell University experimental vineyard in the Cayuga Lake AVA (New York). The experimental vines were pruned to 3 canes per vine with 12 nodes per cane and trained to a Pendelbogen training system (*20*). The experimental site consisted of 3 rows with 20 panels per row and 7 vines per panel.

## Weather Data Collection

Daily temperature and precipitation data from April 1st to Oct 31st were accessed at: http://newa.cornell.edu/index.php?page=all-weather-data. The cumulative growing degree days (GDD) were calculated with respect to 10 °C.

## Sample Collection - Maturity

A panel consisting of 7 vines was treated as a replicate. Four replicate panels were selected from among the 3 rows, and at each sampling time point 300 g of berries were collected from the middle 3 vines of each replicate. In 2009, 8 sampling points were collected at 2, 7, 21, 23, 28, 44, 49, and 56 days post-veraison (dpv). In 2010, 7 sampling points were collected at -7, 2, 7, 23, 42, 49, and 59 dpv. Once collected, samples were kept at -20 °C prior to sample preparation. In 2010 a *Botrytis cinerea* infection started around 42 dpv and affected approximately 50% of fruit by 56 days post-veraison. Only healthy berries, based on visual assessment, were collected from the vines at the sampling points.

## Sample Collection - Hedging

Hedging was performed using a hedge trimmer model C-016 from Echo, Inc. (Lake Zurich, IL). Berries (300 g, in replicate) were sampled immediately prior to hedging and at +2, +7, and +21 days after the hedging treatment. The 6 treatments for 2009 were: 1) Control (no hedging); 2) VO (hedged at 0 days post-veraison); 3) V21 (21 days post-veraison); 4) V42 (42 days post-veraison), 5) V49 (49 days post-veraison); and 6) 2X (hedging at both 0 and 21 days post-veraison). For 2010, the treatments were: 1) Control; 2) V0; 3) V42; 4) Pre-V (-14 days post-veraison); 5) 2X (at 0 and 21 days post-veraison); and 6) 3X (hedging at -14, 21 and 42 days post-veraison). When necessary, shoots were manually tied to stakes to prevent them from shading the fruiting zone.

## Juice Preparation

Using a Waring blender (model no. 5011, Torrington, CT), 200 g of thawed berries were blended, loaded into 85 mL NALGENE polycarbonate centrifuge tubes (VWR International, West Chester, PA), and centrifuged for 30 min at 10,000 rpm and 5 °C (5810 R Centrifuge, VWR International). After centrifuging, the juice supernatant was filtered through No. 41 Whatman filter paper before solid phase extraction (SPE).

## Extraction of Glycosides by SPE

A previously described SPE method for glycoside extraction was adopted (*21*). SPE cartridges (12 mL) were manually packed with 1300 mg of LiChrolut EN sorbent (Merck, Darmstadt, Germany) and processed on a Varian 24-cartridge Positive Pressure Manifold (Palo Alto, CA). Prior to sample loading, the cartridges were pre-conditioned with 32.5 mL dichloromethane, 32.5 mL methanol, and 65 mL $H_2O$. Subsequently, the juice prepared from the 200 g berry sample was loaded onto the SPE cartridge, and the cartridge was washed with 26 mL of $H_2O$ and 40 mL pentane:DCM (2:1 v/v). The retained glycosides were then eluted with 25 mL ethyl acetate:methanol (9:1 v/v). The eluent was first concentrated to ca. 5 mL using a Buchi R-210 Rotavapor at 40 °C and ca. 170 kPa, evaporated to dryness under $N_2$ stream, and finally reconstituted in 10 mL of citric acid buffer (0.2 M, adjusted to pH 2.50).

## Hydrolysis and Extraction of Aglycones

The reconstituted buffer solution was heated in a 100 °C water bath for 1 h in a 20 mL glass vial under a $N_2$-filled headspace. After cooling to room temperature, the 10 mL buffer was spiked with an internal standard mixture, prepared in methanol, to yield final concentrations of 0.2 mg/L 2-octanol for 2009 samples and 0.2 mg/L 4-hydroxy-4-methyl-2-pentanone, 3-ethyl-3-dodecanol, pentanoic acid, 2-octanol, and 2-sec-butyl phenol for 2010 samples. After hydrolysis, the released aglycones (hydrolysate) were isolated by SPE. A pre-packed LiChrolut EN cartridge (200 mg) was preconditioned with 5 mL dichloromethane, 5mL methanol, and 10 mL $H_2O$ before loading the hydrolysate. The cartridge was then dried under a $N_2$ stream (170 kPa for 20 min) and eluted with 2.8 mL dichloromethane. The eluent was then concentrated to ca. 0.3 mL under $N_2$ prior to analysis by gas chromatography mass spectrometry.

## Evaluating Contribution of Non-Glycosylated Compounds to Aglycone Extracts

To evaluate if the water and pentane:DCM (2:1 v/v) washes were effective in removing non-glycosylated compounds, Riesling juice samples were prepared from 56-dpv grapes in 2010, as described above. The SPE-based glycoside extraction was performed on 125 mL samples. Following solvent removal, the sample was reconstituted in citric acid buffer but not heated to minimize acid hydrolysis of glycosides. The reconstituted sample was then immediately extracted by SPE using the same conditions as described for aglycone extractions.

### GC-TOF-MS Analysis of Aglycones

Aglycone extracts were analyzed on a HP6890 GC (Agilent, Santa Clara, CA) coupled to a Pegasus IV time of flight mass spectrometer (Leco, St Joseph, MI). One μL was injected, splitless, onto a DB-Wax column (60 m × 0.25 mm × 0.50 μm, Varian, Walnut Creek, CA) connected to a VF-17 ms (1 m × 0.1 mm × 0.2 μm, Varian). Although the system was set up for GCxGC analyses, the GC×GC modulator was turned off during analysis resulting in 1-D GC-TOF-MS. The injector temperature was set to 250 °C. The purge was opened after 2 min. Helium was used as a carrier gas at a flow rate of 1 mL/min. The temperature program was as follows: initial hold for 1 min at 55 °C, then increased by 3 °C/min to 240 °C with a 60 min final hold. The secondary column and modulator temperature offset was +15 °C. The MS transfer line temperature was 260 °C. The TOF-MS was operated in EI mode with an ionization energy of 70 eV. The electron multiplier was set to 1700 V. MS data from *m/z* 20−400 were stored at an effective sampling rate of 5 Hz.

### Data Processing

Data processing was perfomed by the Leco ChromaTOF® software, similar to an approach described elsewhere (*22*). Initially, the ChromaTOF® software was used to deconvolute and detect peaks (>1000 per analysis) in a representative chromatogram to generate an initial peak list. A final peak list was then generated by retaining only those compounds which could be tentatively identified based on a NIST library version 5.0 search and literature retention indices. Where possible, identification was confirmed by comparison to an authentic standard. The refined peak list, along with the 'unique ion' selected by the software, was copied to a "Reference Table" in the ChromaTOF® software. Then, all files were re-processed using the Reference Table. Peaks were integrated using the previously mentioned 'unique ion' and semi-quantified based on the ratio of the peak area to an internal standard. In 2010, the internal standard mixture consisted of 4-hydroxy-4-methyl-2-pentanone, pentanoic acid, 2-octanol, 3-ethyl-3-dodecanol, and 2-*sec*-butyl-phenol, which were used to quantify ketones, fatty acids, monoterpenes, $C_{13}$-norisoprenoids, and benzenoids respectively. In 2009, only 2-octanol was used as an internal standard.

### Statistical Analysis

Statistical and graphical analyses were performed by JMP version 8 (SAS Institute, Cary, NC).

**Table I. Aglycones Detected by GC-TOF-MS Following Acid Hydrolysis**

| *Quant Mass*[a] | *RI*[b] | *ID*[c] | *Analytes*[d] |
|---|---|---|---|
| | | | **Monoterpenes** |
| 139 | 1131 | A | 2-Ethenyltetrahydro-2,6,6-trimethyl-2H-pyran |
| 93 | 1175 | B | α-Myrcene (2010)* |
| 68 | 1225 | B | Limonene |
| 111 | 1242 | C | Eucalyptol |
| 93 | 1247 | B | cis-Ocimene (2010)* |
| 139 | 1264 | A | Tetrahydro-2,2-dimethyl-5-(1-methyl-1-propenyl)-furan |
| 80 | 1267 | B | trans-Ocimene |
| 93 | 1272 | B | γ-Terpinene |
| 119 | 1298 | B | *p*-Cymene |
| 93 | 1312 | B | Terpinolene |
| 94 | 1473 | B | trans-Linalool oxide, furan |
| 59 | 1502 | B | cis-Linalool oxide, furan |
| 85 | 1502 | B | Nerol oxide |
| 71 | 1554 | A | Linalool (2010)* |
| 93 | 1557 | C | Linalyl acetate |
| 81 | 1604 | B | 1-Terpineol |
| 80 | 1622 | B | Myrcenol |
| 82 | 1624 | B | Hotrienol |
| 111 | 1641 | B | 4-Terpineol |
| 93 | 1661 | B | cis α-Terpineol |
| 93 | 1663 | B | cis-Ocimenol |
| 93 | 1691 | B | trans-Ocimenol |
| 59 | 1732 | C | α-Terpineol |
| 154 | 1733 | C | Terpinyl acetate |
| 68 | 1771 | B | trans-Linalool oxide, pyran |
| 68 | 1787 | B | cis-Linalool oxide, pyran |
| 69 | 1824 | B | Nerol (2010)* |
| 69 | 1867 | A | Geraniol (2010)* |
| 71 | 1984 | B | 6,7-Dihydro-7-hydroxylinalool |
| 81 | 2137 | A | Terpin |

*Continued on next page.*

**Table I. (Continued). Aglycones Detected by GC-TOF-MS Following Acid Hydrolysis**

| *Quant Mass*[a] | *RI*[b] | *ID*[c] | *Analytes*[d] |
|---|---|---|---|
| 81 | 2216 | A | Terpin Hydrate |
| | | | **$C_{13}$-norisoprenoids** |
| 142 | 1781 | C | 1,2-Dihydro-1,1,6-trimethyl-naphthalene |
| 119 | 1581 | B | Vitispirane A + B |
| 121 | 1874 | C | β-Damascenone |
| 163 | 1987 | B | Actinidiol I |
| 163 | 2001 | B | Actinidiol II |
| 157 | 2067 | A | 1,2-Dihydro-1,5,8-trimethyl-naphthalene |
| 132 | 2301 | A | 1-(2,3,6-Trimethylphenyl)-2-butanone |
| 173 | 2368 | A | 1-(2,3,6-Trimethylphenyl)-3-buten-2-one |
| | | | **Benzenoids** |
| 124 | 1904 | C | Guaiacol |
| 108 | 1918 | C | Benzyl alcohol |
| 156 | 1959 | C | Phenylethyl alcohol |
| 150 | 2246 | C | 2-Methoxy-4-vinylphenol |
| 154 | 2313 | C | Syringol |
| 194 | 2369 | B | 2,6-Dimethoxy-4-(2-prepenyl)-phenol |
| 152 | 2628 | C | Vanillin |
| 151 | 2659 | B | Methyl vanillate |
| 151 | 2701 | C | Acetovanilone |
| 137 | 2713 | A | Methyl vanillyl ketone |
| 137 | 2845 | C | Gingerone |
| 137 | 2891 | C | Homovanillyl alcohol |
| 182 | 2987 | C | Syringaaldehyde |
| 137 | 3025 | C | Homovanilic acid |

[a] Quant mass was the automatically assigned unique mass by Leco deconvolution software. [b] Retention index on DB-Wax column. [c] Identification was performed by library search (a) with literature retention index (b) and standard verification (c). [d] The five monoterpenes odorants detected only in the 2010 season are indicated with an asterisk (*).

## Results and Discussion

### Weather Data and Basic Juice Chemistry

The 2010 growing season was warmer than the 2009 growing season (1551 GDD vs. 1278 GDD), with most of the difference in GDD occurring pre-veraison. The 2010 growing season was also wetter than the 2009 growing season (595 mm vs 439 mm), with most of the difference between the years occurring post-veraison. In summary, 2010 was warmer pre-veraison and wetter post-veraison than 2009. The basic juice chemistry at 56 dpv, the final harvest point, for 2009 was as follows: soluble solids = 21.1 Brix, TA = 10.7 g/L as tartaric, and pH = 3.06. In 2010, the basic juice chemistry was as follows: soluble solids = 21.3 Brix, TA = 7.2 g/L, and pH = 3.04.

### Glycoside Profiles

Acid hydrolysis was selected instead of enzymatic hydrolysis because several volatiles of interest in Riesling wine require acid catalyzed rearrangements following hydrolysis of the glycoconjugate, e.g., vitispirane, β-damascenone, and 1,1,6-trimethyl-1,2-dihydronaphthalene (TDN). A list of compounds measured including their means of identification is presented in Table I. The volatiles detected included monoterpenes, $C_{13}$-norisoprenoids, and benzenoids. The listed compounds have been previously observed in the acid-hydrolyzed glycoside fractions extracted from grapes (*6*, *23*). Five monoterpenes (linalool, nerol, geraniol, *cis*-ocimene, and α-myrcene) that were not detected in 2009 were detected in 2010 samples. Other groups have noted that acid hydrolysis leads to reduced recovery of linalool, nerol and geraniol (*8*), as these monoterpenes are known to rearrange under acidic conditions (*24*).

### Identifying Incompletely Removed "Free" Volatiles by Comparing Hydrolyzed and Non-Hydrolyzed Samples

Several volatiles listed in Table I, such as the aldehydes and fatty acids, seemed unlikely to exist as glycosides because they lack an –OH group, and these compounds were suspected of being incompletely removed during cartridge washing. To evaluate if apparently 'bound' volatiles were in fact free volatiles that were incompletely removed during the SPE wash step in the SPE protocol for glycoside extraction, we repeated the SPE protocol on the original juice samples without the hydrolysis step. If the wash step was efficient, then minimal signal should have been observed in the non-hydrolyzed sample. Of the 67 aglycones reported in Table I in the hydrolyzed sample, only 20 analytes in the non-hydrolyzed sample had ≥1% of the signal observed in the hydrolyzed sample (Table II). These compounds were mostly short chain or aromatic fatty acids, alcohols, and aldehydes that have been previously reported in studies of free volatiles in grapes (*12*, *13*). Monoterpenes and $C_{13}$-norisoprenoids were not generally observed in the non-hydrolyzed sample. One notable exception was 3,7-dimethyl-1,5-octadiene-3,7-diol, which had a signal over 3-fold greater (364%) in the non-hydrolyzed sample than the hydrolyzed sample. This compound, referred as diendiol I in some reports (*9*), is present in a free form in grapes and can degrade to form other monoterpene alcohols under acidic conditions. The apparent decrease in diendiol I in the hydrolysis sample likely indicates that diendiol I is incompletely removed by the SPE wash step and degraded during thermal acid treatment. To avoid these artifacts, we excluded compounds with peak area ratios >26% from further consideration. (*E*)-2-hexenal was also excluded, since it seemed unlikely to be derived from a glycoside.

**Table II. Ratio of Volatile Peak Areas from Non-Hydrolyzed and Hydrolyzed Samples for Compounds with Ratio >1%**

| *Name* | *Peak area ratio* * |
|---|---|
| (*E*)-2-Hexenal | 16% |
| 1-Hexanol | 47% |
| (*E*)-2-Hexen-1-ol | 64% |
| 2-Methyl-propanoic acid | 110% |
| Butanoic acid | 62% |
| Hexanoic acid | 60% |
| (*E*)-2-Hexenoic acid | 35% |
| 3,7-Dimethyl-1,5-octadiene-3,7-diol | 364% |
| Hotrienol | 3% |
| Benzyl alcohol | 11% |
| Phenylethyl alcohol | 9% |
| Benzaldehyde | 12% |
| Butyrolactone | 1% |
| Pantolactone | 1% |
| Phenol | 51% |
| *p*-Cresol | 25% |
| 2-Methoxy-4-vinylphenol | 1% |
| Vanillin | 9% |
| Homovanillyl alcohol | 26% |
| Homovanilic acid | 23% |

* Percent of free fraction retained on SPE column was calculated by the peak ratio of non-hydrolyzed sample divided by its hydrolyzed sample.

## Behavior of Individual Aglycones during Berry Ripening

For each compound, the peak area at each time point was normalized to the maximum value observed across all time points for the compound. These mean normalized values, n=4 biological replicates, are depicted as a heat map in Figure 1. In accordance with previous studies, we observed that most monoterpene (*9*, *10*, *25*, *26*) and $C_{13}$-norisoprenoid (*11*, *25*, *27*) aglycones increased during ripening in both growing seasons. For example, bound monoterpenes such as α-terpineol, nerol, geraniol, and linalool showed an increase with maturation in our current study and others (*10*, *26*). We did, however, observe a ca. 10-fold increase for α-terpineol between veraison and harvest in our current study, compared to the 2-fold increase reported in other studies that relied on enzymatic hydrolysis (*10*, *26*). Other authors have noted that acid hydrolysis can yield higher concentration of α-terpineol than enzymatic hydrolysis, especially for high monoterpene cultivars (*8*). Thus, it is likely that some portion of apparently bound α-terpineol in our current work is derived from other monoterpene glycosides.

Concentrations of all observed bound $C_{13}$-norisoprenoids, such as 1,1,6-trimethyl-1,2-dihydronaphthalene (TDN), vitispirane and β-damascenone, also increased during berry maturation. Since the $C_{13}$-norisoprenoid aglycones formed during enzymatic and acid hydrolyses are qualitatively different (*8*, *28*, *29*), it is not possible to compare the behavior of specific $C_{13}$-norisoprenoids in this study to previous reports that utilized enzymatic hydrolysis (*11*, *27*). One exception which utilized acid hydrolysis is a report on $C_{13}$-norisoprenoids in South African Riesling (*30*), which observed increases in TDN and β-damascenone during ripening. However, the earliest sampling points in this work were at 16-18 Brix, presumably several weeks after veraison. The behavior of benzenoid aglycones was inconsistent across the two growing seasons (data not shown). In 2009, most benzenoids (9 of 14) reached their maximum value at 28 dpv before decreasing, while 10 of 14 reached their maximum at 56 dpv in 2010. Homovanillyl alcohol and syringaldehyde increased throughout ripening in both seasons, but other phenols like 2-methoxy-4-vinylphenol, syringol, and 2,6-dimethoxy-4-(2-propenyl)-phenol increased during ripening only in 2010. Other benzenoids, like guaiacol in 2009, were at a maximum near veraison and then decreased during ripening. Time course studies on bound benzyl alcohol and phenylethanol have been previously reported (*31*) and do not appear to show a consistent pattern. Simulated "smoke taint" experiments have demonstrated that smoke exposure around veraison leads to an increase in glycosylated volatile phenols such as guaiacol, with lower concentrations observed for earlier or later smoke exposure (*32*). However, only a single time point (harvest) was studied in these experiments, and it is not clear if final differences result from differences in mass transfer or metabolic activity.

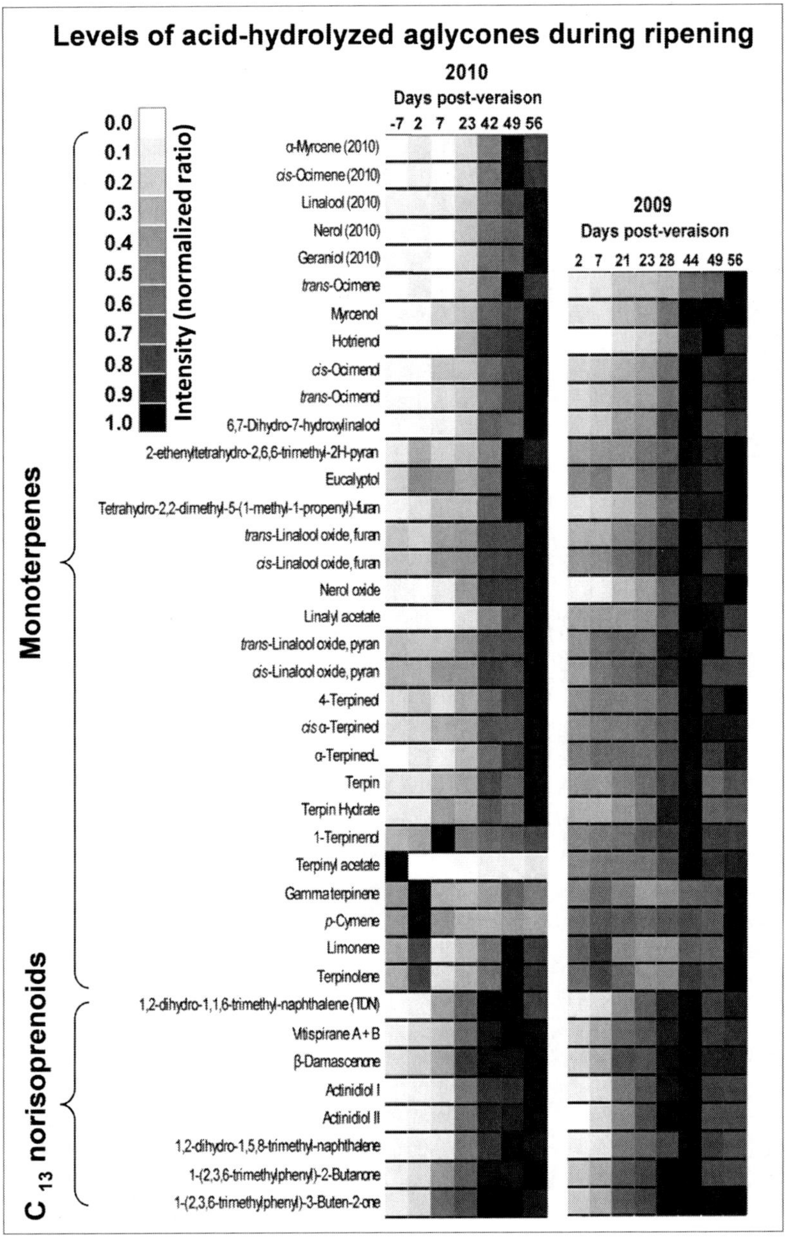


*Figure 1. Concentrations of acid-hydrolyzed aglycones during ripening in the 2009 and 2010 seasons, normalized against the maximum concentration observed for each compound.*

## Within Class Pairwise Correlations for Oxygenated Monoterpenes and $C_{13}$-Norisoprenoids

Based on pairwise correlations, we observed that all monoterpenes were strongly and positively correlated with each other (r>0.9) with the exception of eucalyptol, which had a correlation of >0.8 with several monoterpenes. The strong correlation among monoterpenes was not unexpected, since the compounds not only share common biochemical origins but also may be partially interconverted during winemaking (*33*). $C_{13}$-norisoprenoid aglycones were also highly correlated with each other (r>0.9), similar to a previous report (*34*). Considering the strong correlations of aglycones within each class, measuring a subset of aglycones may be sufficient for studying the behavior of a particular aglycone class.

## Differences in Timing of Monoterpene and $C_{13}$-Norisoprenoid Glycosides Accumulation during Ripening

The data in Figure 1 suggest that the majority of $C_{13}$-norisoprenoid glycoside accumulation occurs before the majority of monoterpene glycoside accumulation. To further evaluate this hypothesis, the mean normalized concentrations at each time point for each compound class were compared using a Tukey test. Plots of the mean normalized concentrations for both growing seasons are shown in Figure 2. Although both classes increased after veraison in both years, we observed that monoterpene glycosides started accumulation later and reached their maximum later than $C_{13}$-norisoprenoid glycosides. We observed a significant increase in bound $C_{13}$-norisoprenoids by 21 dpv in 2009 and by 7 dpv in 2010. Significant increases were not observed. No significant accumulation ($p>0.05$, ANOVA) was noted for monoterpene glycosides until 23 dpv. These results suggest that accumulation of bound $C_{13}$-norisoprenoids starts earlier and peaks prior to monoterpene glycoside accumulation. One potential explanation is that d*e novo* monoterpene synthesis appears to continue during berry ripening and can then serve as a substrate for the production of monoterpene glycosides. Previous studies on monoterpene glycosides in Muscat varieties have shown that bound monoterpenes will start to accumulate within one week post-veraison and will continue to accumulate for up to 60 dpv (*3*, *10*, *26*). In contrast, $^{13}C$ tracer studies indicate that the first step in $C_{13}$-norisoprenoid glycoside production occurs through degradation of carotenoids during veraison (*27*). In this situation, bound forms of $C_{13}$-norisoprenoids are expected to plateau soon after veraison once the carotenoid substrate is consumed, possibly by a carotenoid cleavage dioxygenase enzyme (*11*). Previous work observed that accumulation of $C_{13}$-norisoprenoids starts one week after veraison, but because this earlier work only sampled berries until 4.5 weeks post-veraison, it is not possible to evaluate if a plateau occurs. Other studies suggest that $C_{13}$-norisoprenoid glycosides can increase dramatically after 40 dpv (*27*, *35*), although no statistical analyses were provided in these reports.

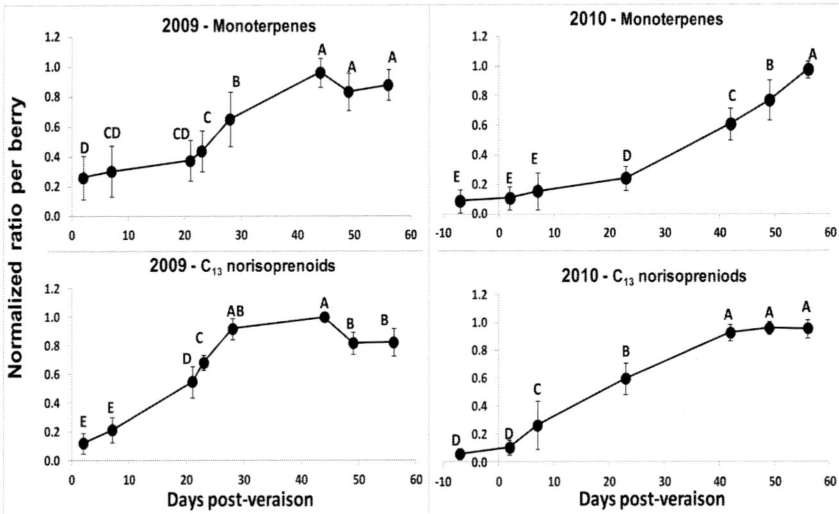


*Figure 2. Behavior of monoterpene (top) and $C_{13}$ norisoprenoid (bottom) aglycones during grape ripening in the 2009 (left) and 2010 (right) seasons. The y-axis represents the mean value of the normalized ratios within an aglycone class from Figure 1, excluding those compounds that appear to be extraction artifacts. Different letters indicate significant differences, determined by a Tukey test ($p<0.05$).*

## Effects of Hedging Treatments on Glycoside Accumulation

Because cluster light exposure has been previously associated with monoterpene accumulation (*20*), we made efforts to ensure that the hedging treatment did not alter the amount of sunlight reaching the grape berries. Grape shoots that were not hedged, e.g., in the control treatment, were tied up to prevent them from shading clusters. We also confirmed that the fruit was well exposed by performing Enhanced Point Quadrat Analysis (*36*) and observed no differences in any of the metrics associated with cluster exposure (data not shown).

Across both years, we observed no significant differences in the normalized concentrations of the different compound classes as a result of any of the treatments (ANOVA, $p>0.05$). We did observe small (<30%) but significant increases ($p<0.05$) for vitispirane, 1,2-dihydro-1,5,8-trimethyl-napthalene, 1-(2,3,6-trimethylphenyl)-2-butanone, and syringol in 2009 and guaiacol in 2010 for the V42 and V56 treatments. However, these trends were not consistent across years. Furthermore, the effects were small compared to the effects of other well known viticultural practices, i.e., cluster light exposure can result in a >2 fold increase in some $C_{13}$ norisoprenoids (*19*).

## Conclusions

Although most monoterpenes and $C_{13}$-norisoprenoids glycosides exhibit a consistent pattern of increasing during grape ripening, the timing of their accumulation is different, as $C_{13}$-norisoprenoid glycosides accumulate and plateau approximately one to two weeks before monoterpenes do. This suggests that harvest time may have a critical impact on the odor character of wines like Riesling which can have both monoterpenes (linalool, geraniol) and $C_{13}$-norisoprenoids (TDN) at suprathreshold concentrations. Within its aroma class, most monoterpene and $C_{13}$-norisoprenoid aglycones were highly correlated ($r>0.9$), suggesting that it should be possible to measure representative compounds within each class. Finally, late-season hedging does not result in a clear increase in glycosides in grapes at harvest. It may be of future interest to use chemical elicitors of grape defense pathways to further explore this matter, as these compounds have been shown to increase production of some phenolics (*17*) as well as S-cysteine conjugates (*18*).

## References


1. Baumes, R. In *Wine Chemistry and Biochemistry*; Moreno-Arribas, M. V., Polo, M. C., Eds.; Springer: New York, 2009; pp 251−274.
2. Ribéreau-Gayon, P.; Glories, Y.; Maujean, A.; Dubourdieu, D. *Handbook of Enology, Vol.2, The Chemistry of Wine Stabilization and Treatments*, 2nd ed.; John Wiley & Sons: Chichester, 2006.
3. Park, S. K.; Morrison, J. C.; Adams, D. O.; Noble, A. C. *J. Agric. Food Chem.* **1991**, *39*, 514–518.
4. Ugliano, M.; Bartowsky, E. J.; McCarthy, J.; Moio, L.; Henschke, P. A. *J. Agric. Food Chem.* **2006**, *54*, 6322–6331.
5. Winterhalter, P.; Sefton, M. A.; Williams, P. J. *Am. J. Enol. Vitic.* **1990**, *41*, 277–283.
6. Loscos, N.; Hernández-Orte, P.; Cacho, J.; Ferreira, V. *Food Chem.* **2010**, *120*, 205–216.
7. Mateo, J. J.; Jiménez, M. *J. Chromatogr., A* **2000**, *881*, 557–567.
8. Loscos, N.; Hernández-Orte, P.; Cacho, J.; Ferreira, V. *J. Agric. Food Chem.* **2009**, *57*, 2468–2480.
9. Wilson, B.; Strauss, C. R.; Williams, P. J. *J. Agric. Food Chem.* **1984**, *32*, 919–924.
10. Fenoll, J.; Manso, A.; Hellín, P.; Ruiz, L.; Flores, P. *Food Chem.* **2009**, *114*, 420–428.
11. Mathieu, S.; Terrier, N.; Procureur, J.; Bigey, F.; Gunata, Z. *J. Exp. Bot.* **2005**, *56*, 2721–2731.
12. Kalua, C. M.; Boss, P. K. *J. Agric. Food Chem.* **2009**, *57*, 3818–3830.
13. Canuti, V.; Conversano, M.; Calzi, M. L.; Heymann, H.; Matthews, M. A.; Ebeler, S. E. *J. Chromatogr., A* **2009**, *1216*, 3012–3022.
14. Kennison, K. R.; Gibberd, M. R.; Pollnitz, A. P.; Wilkinson, K. L. *J. Agric. Food Chem.* **2008**, *56*, 7379–7833.

15. Cha, D. H.; Nojima, S.; Hesler, S. P.; Zhang, A.; Linn, C. E.; Roelofs, W. L.; Loeb, G. M. *J. Chem. Ecol.* **2008**, *34*, 1180–1189.
16. Kant, M. R.; Bleeker, P. M.; Van Wijk, M.; Schuurink, R. C.; Haring, M. A. In *Plant Innate Immunity*; Van Loon, L. C., Ed.; Academic Press Ltd-Elsevier Science Ltd: London, 2009; Vol. 51, pp 613−666.
17. Ruiz-García, Y.; Romero-Cascales, I.; Gil-Muñoz, R.; Fernández-Fernández, J. I.; López-Roca, J. M.; Gómez-Plaza, E. *J. Agric. Food Chem.* **2012**, *60*, 1283–1290.
18. Kobayashi, H.; Takase, H.; Suzuki, Y.; Tanzawa, F.; Takata, R.; Fujita, K.; Kohno, M.; Mochizuki, M.; Suzuki, S.; Konno, T. *J. Exp. Bot.* **2011**, *62*, 1325–1336.
19. Kwasniewski, M. T.; Vanden Heuvel, J. E.; Pan, B. S.; Sacks, G. L. *J. Agric. Food Chem.* **2010**, *58*, 6841–6849.
20. Reynolds, A. G.; Vanden Heuvel, J. E. *Am. J. Enol. Vitic.* **2009**, *60*, 251–268.
21. Jesús Ibarz, M.; Ferreira, V.; Hernandez-Orte, P.; Loscos, N.; Cacho, J. *J. Chromatogr., A* **2006**, *1116*, 217–229.
22. Johanningsmeier, S. D.; McFeeters, R. F. *J. Food Sci.* **2011**, *76*, C168–C177.
23. Sefton, M. A.; Francis, I. L.; Williams, P. J. *Am. J. Enol. Vitic.* **1993**, *44*, 359–370.
24. Gunata, Y. Z.; Bayonove, C. L.; Baumes, R. L.; Cordonnier, R. E. *Am. J. Enol. Vitic.* **1986**, *37*, 112–114.
25. Razungles, A.; Gunata, Z.; Pinatel, S.; Baumes, R.; Bayonove, C. *Sci. Aliments* **1993**, *13*, 59–72.
26. Gunata, Y. Z.; Bayonove, C. L.; Baumes, R. L.; Cordonnier, R. E. *J. Sci. Food Agric.* **1985**, *36*, 857–862.
27. Baumes, R.; Wirth, J.; Bureau, S.; Gunata, Y.; Razungles, A. *Anal. Chim. Acta* **2002**, *458*, 3–14.
28. Schneider, R.; Razungles, A.; Augier, C.; Baumes, R. *J. Chromatogr., A* **2001**, *936*, 145–157.
29. De Rosso, M.; Panighel, A.; Carraro, R.; Padoan, E.; Favaro, A.; Dalla Vedova, A.; Flamini, R. *J. Agric. Food Chem.* **2010**, *58*, 11364–11371.
30. Marais, J.; Van Wyk, C. J.; Rapp, A. *S. Afr. J. Enol. Vitic.* **1992**, *13*, 23–32.
31. Hellín, P.; Manso, A.; Flores, P.; Fenoll, J. *J. Agric. Food Chem.* **2010**, *58*, 6334–6340.
32. Kennison, K. R.; Wilkinson, K. L.; Pollnitz, A. P.; Williams, H. G.; Gibberd, M. R. *Aust. J. Grape Wine Res.* **2011**, *17*, S5–S12.
33. Margalit, Y. *Concepts in Wine Chemistry*, 2nd ed.; Wine Appreciation Guild: San Francisco, CA, 2004.
34. Strauss, C. R.; Wilson, B.; Anderson, R.; Williams, P. J. *Am. J. Enol. Vitic.* **1987**, *38*, 23–27.
35. Marais, J.; Versini, G.; Van Wyk, C. J.; Rapp, A. *S. Afr. J. Enol. Vitic.* **1992**, *13*, 71–77.
36. Meyers, J. M.; Vanden Heuvel, J. E. *Am. J. Enol. Vitic.* **2008**, *59*, 425–431.

## Chapter 11

# TDN and β-Damascenone: Two Important Carotenoid Metabolites in Wine

**Peter Winterhalter*** **and Recep Gök**

**Institut für Lebensmittelchemie, Technische Universität Braunschweig, Schleinitzstraße 20, 38106 Braunschweig, Germany**
***E-mail: p.winterhalter@tu-bs.de.**

This chapter presents an overview on the generation of the norisoprenoids 1,1,6-trimethyl-1,2-dihydronaphthalene (TDN) and β-damascenone in wine. TDN is responsible for a "kerosene note" in aged Riesling wines. Responsible for TDN formation are a number of reactive nonvolatile carotenoid cleavage products. For TDN formation warmer climates and increased sunlight exposure of the grapes are known to have a considerable impact. Measures for avoiding this off-flavor during wine production are discussed. With regard to the formation of the odoriferous β-damascenone a series of acid-labile acetylenic and allenic precursors has been identified revealing a complex pathway of formation for β-damascenone in wine.

Carotenoid-derived metabolites are important aroma constituents in grapes and wines (Figure 1). This is due to their generally low flavor threshold which in the case of β-damascenone **2** and β-ionone **3** is in the ppb-range (*1*). The principle steps of bio-oxidative carotenoid-degradation have been elucidated in recent years (*2*, *3*), however there is still a considerable lack of knowledge especially with regard to the enzymatic conversions of the primary carotenoid cleavage products (Step II, cf. Figure 2) giving rise to the formation of reactive aroma precursors.

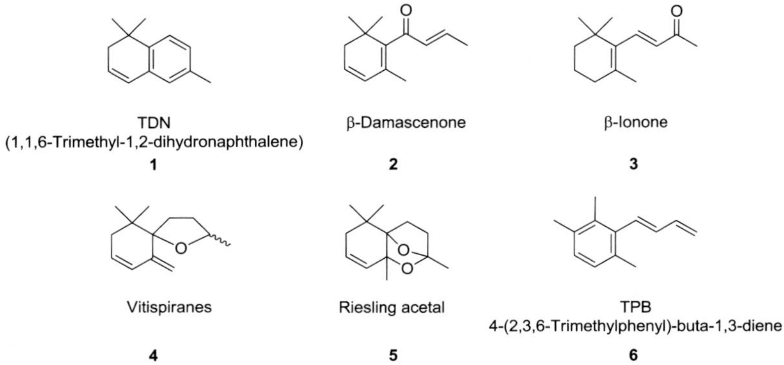


*Figure 1. Examples for carotenoid-derived wine aroma compounds.*

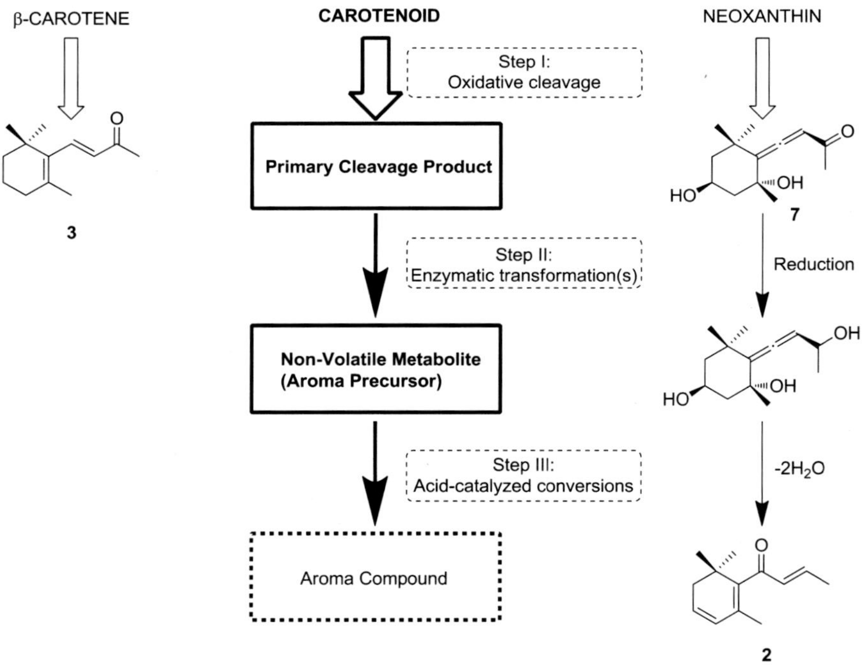


*Figure 2. Bio-oxidative carotenoid cleavage giving rise to primary degradation products and the subsequent (bio)chemical conversion into aroma compounds.*

These primary cleavage products are obtained by the action of a carotenoid cleavage dioxygenase (CCD) which after cleavage at the C9,10 double bond liberates the endgroup consisting of 13 carbon atoms ($C_{13}$-norisoprenoid) and a $C_{27}$-apocarotenoid. The best studied reaction is the CCD-catalyzed formation of β-ionone **3** from β-carotene ((*3*), and refs. cited therein). Another example is the cleavage of neoxanthin to grasshopper ketone **7** which is an important progenitor of β-damascenone **2**. Less information is available for the respective initial cleavage products which are involved in TDN formation. From biogenetic considerations 3-hydroxy-5,6-epoxy-β-ionone **8** being either derived from violaxanthin or neoxanthin is the most likely candidate, cf. Figure 3 (*4*). However, since labeling experiments are missing, pathway of TDN formation still requires more thorough examination.

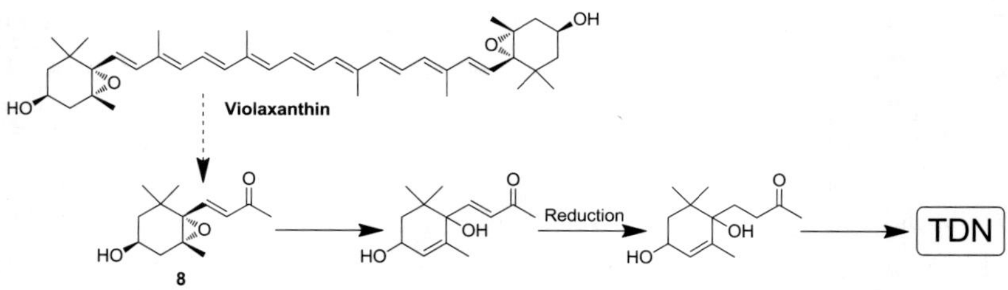


*Figure 3. Postulated formation of TDN **1** from 3-hydroxy-5,6-epoxy-β-ionone **8**.*

## Formation of TDN

TDN **1** is known as an important contributor to the bottle-aged bouqet of Riesling wine. It has first been identified in 1978 by Simpson (*5*) who determined its flavor threshold in wine in the range of 20 µg/L. In the meantime, the flavor threshold of TDN has been redetermined by Sacks and coworkers (*6*) revealing a threshold a factor of ten below the previously reported one, i.e. 2 µg/L. When present in higher concentration, TDN is responsible for an off-flavor that is decribed as "*kerosene*" or "*petrol-like*". Mainly effected are Riesling grapes/wines from the Southern hemisphere, however in the course of global warming also the European mainland is expected to have increasing problems with TDN. Reasons are the well-documented higher levels of TDN for grapes grown under warmer climatic conditions and/or under higher sun-light exposure (*7*, *8*, *35*). In this context it is important to note that mainly the nonvolatile progenitors of TDN increase in concentration. Hence measures for delaying or reducing the TDN off-flavor is either to keep these progenitors intact during the storage of wine (e.g. by cool storage) or to transform them into less harmful analogs. This aspect will be discussed later.

### Synthesis of TDN and Deuterium-Labeled TDN

In order to get an overview on TDN levels in German Riesling wines a three-years survey employing a stable isotope dilution assay (SIDA) is presently carried out. There are only few methods described in the literature for the synthesis of TDN. These methods are based on an intramolecular cyclization of α-

or β-ionone to ionene **9** and the subsequent conversion of **9** to TDN **1**. In addition also a "direct" synthesis of TDN has been reported because TDN is obtained as by-product after treatment of α- or β-ionone with N-bromo-succinimide followed by dehydrobromination with *N,N*-diethylaniline (*9*, *10*). The most common method for the synthesis of TDN is a two-step reaction first published by Miginiac (*11*). This method consists of heating α-ionone with iodine and treatment of the product (i.e. ionene) with *N*-bromosuccinimide (NBS) and *N,N*-diethylaniline (*11*). Another strategy used in our experiments is an ultrasound-mediated rearrangement of β-ionone **3** to ionene **9** and the subsequent introduction of the double-bond by treatment of ionene with NBS and *N, N*-diethylaniline (*12*). Importantly, these methods yield a mixture of TDN **1** and ionene **9** and further purification by e.g. high speed countercurrent chromatography (HSCCC) is necessary. HSCCC was operated in the reversed-phase elution mode (head-to-tail) employing the nonaqueous solvent system *n*-hexane/acetonitrile 1:1 (v/v). The separation was performed on a multilayer coil planet centrifuge model CCC-1000 (Pharma-Tech Research Corp.; Baltimore, MD, USA) equipped with three preparative coils connected in series (total volume: 850 mL, flow rate: 3.0 mL/min, rotational speed: 1000 rpm). Fractions were monitored by TLC, visualized under UV light (λ 254 nm) and the purity of the fractions were determined by GC/MS analysis (Figure 4).

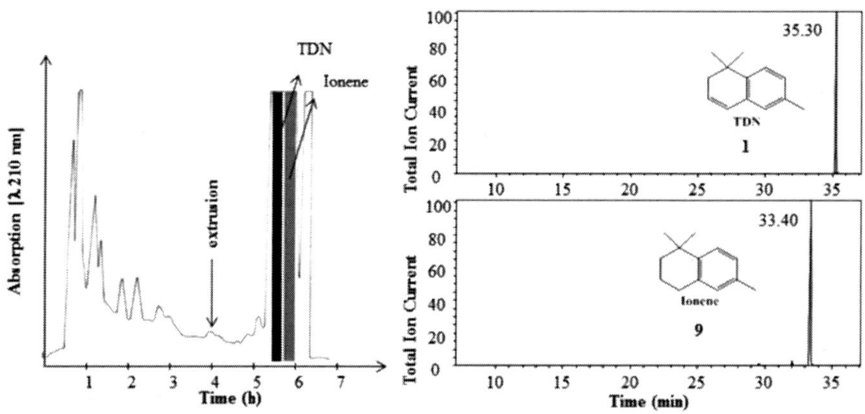


*Figure 4. HSCCC chromatogram of the reaction mixture of TDN and ionene and the GC/MS chromatograms of the purified compounds.*

For the synthesis of a deuterated TDN sample [$^2H_3$]-β-ionone (*13*) has been used as an educt (Figure 5). The survey on TDN levels in German wines is still in progress and will be published separately. In general TDN levels were in the low ppb range which is in accordance with published data from the United States (*6*).

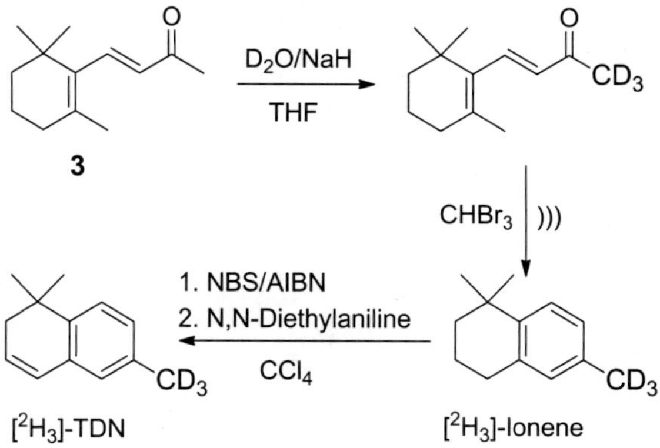


*Figure 5. Synthesis of deuterated TDN (for details cf. text).*

## Formation of TDN from Nonvolatile Precursors

TDN **1** is known to be generated from nonvolatile progenitors under acidic conditions. The occurrence of TDN is not restricted to wine and grapes, hence different progenitors for TDN were detected in different plants. In the juice of purple passionfruit (*Passiflora edulis* Sims) a glycoconjugate of 3-hydroxy-1,1,6-trimethyl-1,2,3,4-tetrahydronaphthalene **10** could be identified (*14*), whereas in quince fruit (*Cydonia oblonga* Mill.) formation of TDN was due to the breakdown of glycosidically bound 3-hydroxy-β-ionone **11** (*15*).

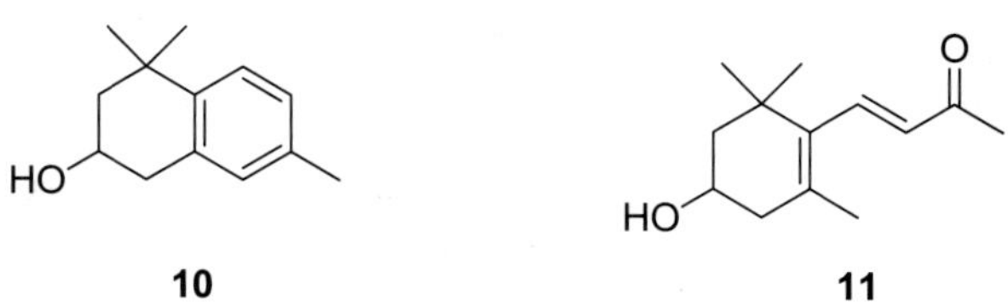


The breakthrough in TDN research in wine dates back to the year 1991 when a series of acid-labile TDN progenitors has been detected in Riesling wine (*16*). Key progenitors for TDN in wine are 3,6-dihydroxy-7,8-dihydro-α-ionone **12** being present in form of the diastereomeric hemiacetals **12a/b** as well as the allylic rearranged dihydroxyketone **13**. These carotenoid-derived metabolites are present in multiple glycoconjugated forms in wine and the aglycones **12a/b** and **13** were released after glycosidase treatment (*16*). The complete structural characterization of the intact glycoconjugates with regard to the sugar moieties employed still remains to be carried out.

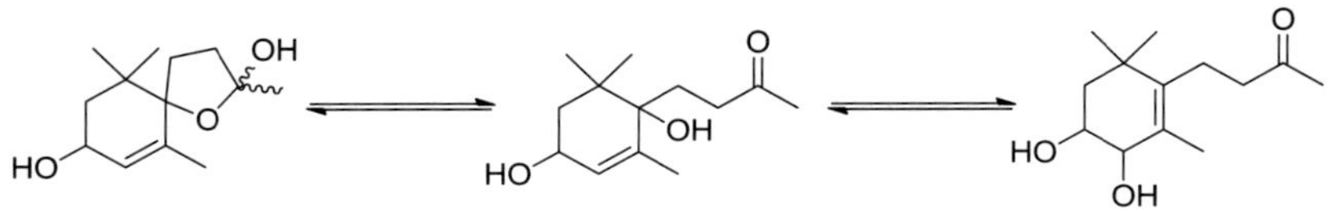


Under acidic conditions the reactive progenitors **12, 12a/b** and **13** are converted into a series of volatile compounds, which are shown in Figure 6.

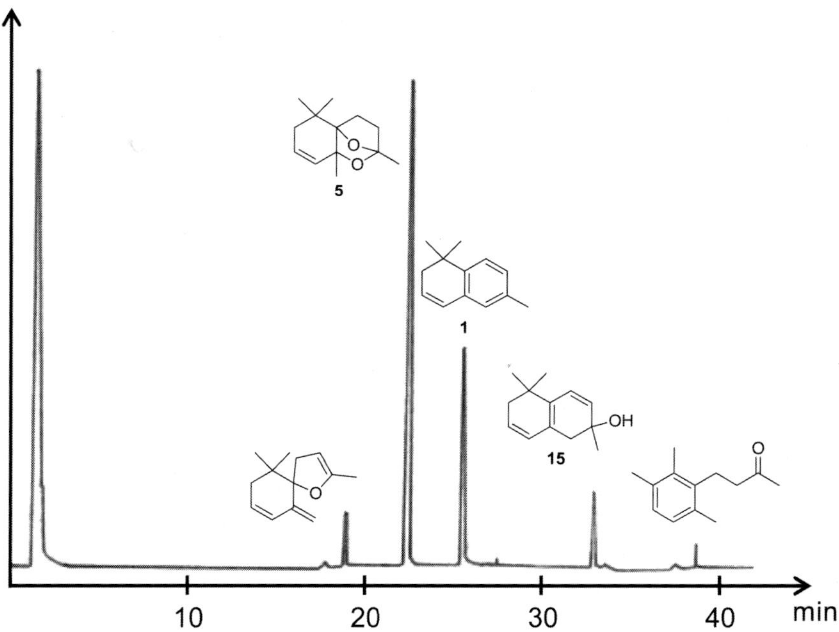


*Figure 6. Volatile reaction products obtained from progenitors **12a/b** and **13**. (adapted from ref. (16)).*

As major degradation products Riesling acetal **5** (which eventually will also generate TDN (*17*)) and the target compound TDN **1** are obtained. The same pattern of volatiles has been detected in separated precursor fractions of Riesling wine except for one fraction in which (upon heating) only TDN **1** and hydroxy-TDN **15** is formed. This indicates that in addition to the known progenitors **12a/b** and **13** at least one additional precursor awaits to be elucidated. The most likely candidate is the hydroxyketone **14** which upon dehydration and cyclization reactions is expected to give rise to TDN **1** and the hydroxylated intermediate **15**.

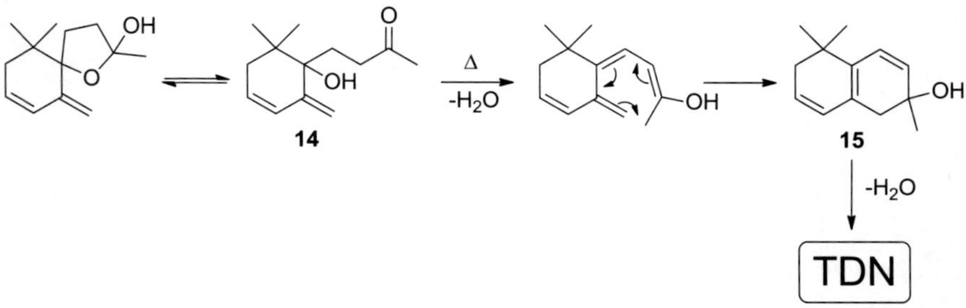


The reduced form of **14**, i.e. 1-(3-hydroxybutyl)-6,6-dimethyl-2-methylene-3-cyclohexen-1-ol **16,** is known to be present in Riesling in glycoconjugated form. Diol **16** also has an aroma precursor function. Under acidic conditions diol **16** is converted into isomeric vitispiranes **4** (*18*).

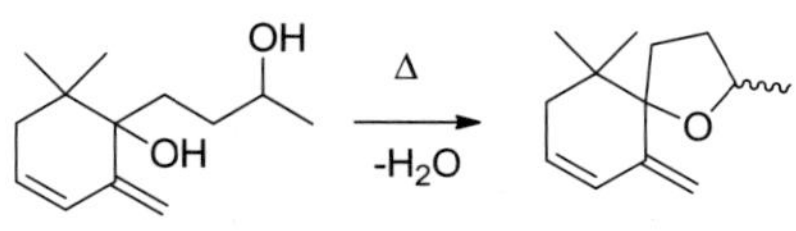


**16** **4**

This brings us to an important issue of TDN research, namely possible measures for avoiding the development of this off-flavor.

## Prevention of TDN Taint

Due to climatic changes (warmer climate, higher sunlight exposure) Riesling grapes are expected to develop increasing levels of TDN progenitors, and hence Riesling wines will most likely be more prone to the development of a *kerosene* off-flavor during storage. The reasons behind the detected higher levels of TDN progenitors especially in sun-exposed Riesling grapes are still not clear and more information about enzymatic and nonenzymatic carotenoid breakdown in Riesling grapes under high sunlight conditions is required. Some speculations are centered around the so-called violaxanthin cycle, which is used by plants to adapt to different light conditions and allows a reversible switch of photosynthetic light-harvesting complexes between a light-harvesting state under low light and a dissipative state under high light. In the course of this cycle, violaxanthin is deepoxidized to zeaxanthin via the intermediate anther-xanthin (*12*). There is also evidence for the functioning of the lutein epoxide cycle in grapevine (*36*). What distinguishes Riesling from the other varieties in this regard and what causes higher breakdown of the carotenoid precursors in this variety are still questions that remain to be answered .

Nevertheless measures are available for reducing the risk of an off-flavor development. Apart from cool storage of the wines, which prevents/retards the conversion of the progenitors into the off-flavor compound TDN, one strategy involves the reduction of the ketone function at carbon atom 9 in the glycosidically bound TDN progenitors, i.e **3-O-Glc-12, -12a/b** and **-13** (cf. Figure 7). Model reactions carried out with bakers yeast enabled a substantial but not complete reduction of TDN generating compounds (*19*) as could be seen from reduced levels of TDN and enhanced levels of isomeric vitispiranes **4**. At present a screening is carried out which is expected to shed more light on the potential of wine yeasts in reducing TDN progenitors into the less harmful glycosidically bound vitispirane progenitors **9-O-Glc-16, 3-O-Glc-17,** and **3-O-Glc-18**. This approach is seen as a feasible tool to minimize TDN taints during wine processing and storage.

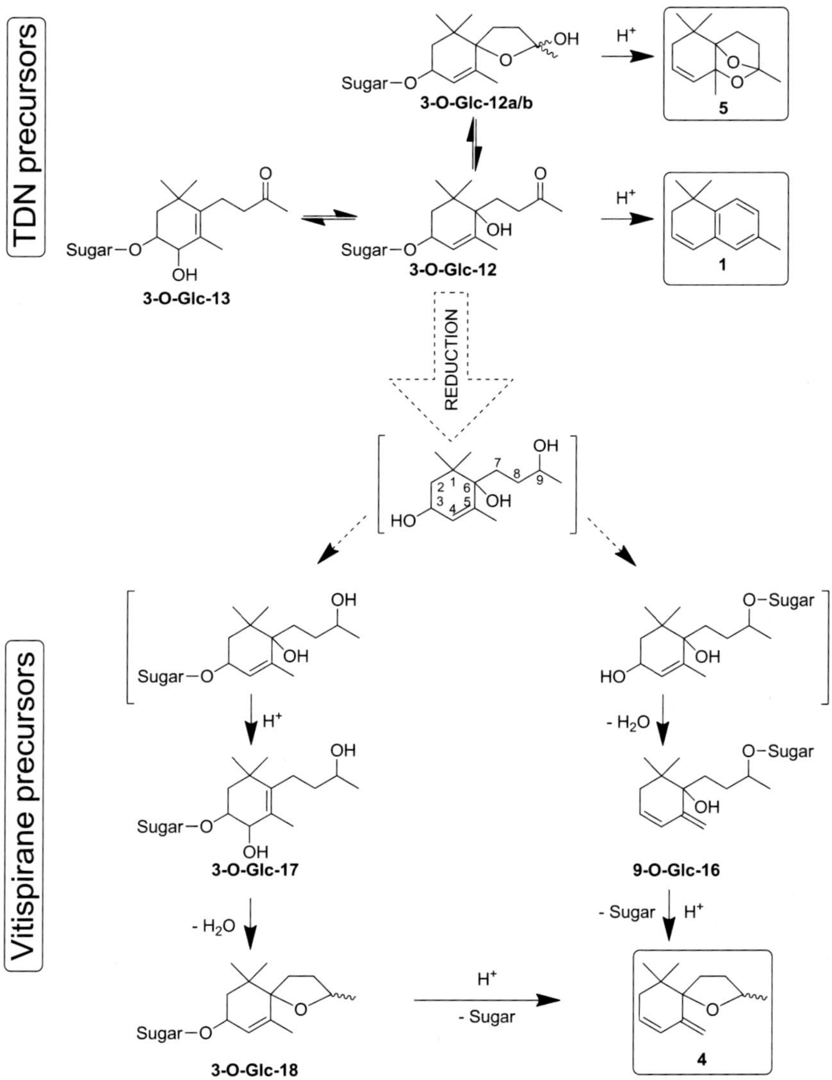


*Figure 7. Wine yeast mediated reduction of TDN progenitors **3-O-Glc-12/-12a/b** and **-13** into vitispirane precursors **9-O-Glc-16, 3-O-Glc-17** and **-18** as a measure to reduce the risk for TDN off-flavors in wine.*

## Formation of β-Damascenone

The second important carotenoid-derived aroma compound in wine is β-damascenone **2**. This odoriferous ketone is ubiquitous in nature and its generation from carotenoids has been independently proposed almost 40 years ago by Isoe *et al.* (*20*) and Ohloff *et al.* (*21*). Neoxanthin has been suggested as likely carotenoid precursor and formation of β-damascenone was considered to take place via grasshopper ketone **7**, allenic triol **19**, and 3-hydroxy-β-damascone **20** (cf. Figure 8). Today we know that this hypothesis was almost correct except that from the allenic triol **19** different intermediates are involved in the formation of the key flavor compound **2**. Moreover, hydroxyketone **20** is stable under pH conditions of wine and will not be converted into the target compound β-damascenone **2**. A comprehensive review on the occurrence, sensory impact and formation of β-damascenone **2** in food has recently been published by Sefton *et al.* (*22*), summarizing the published literature about β-damascenone until the year 2010. In the following only selected aspects of β-damascenone formation in wine will be presented.

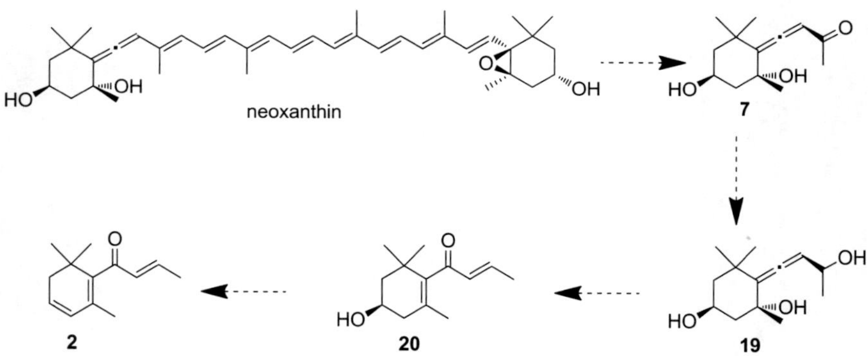


*Figure 8. Proposed pathway of β-damascenone formation. Adapted with permission from Reference (22). Copyright 2011, American Chemical Society.*

The isolation of intact β-damascenone precursors from wine medium is difficult to achieve because of their low concentrations in wine as the well as the rather low pH of wine (pH 3.2) which causes acid-catalyzed degradations of the extremely reactive aroma progenitors. By application of the gentle isolation technique of multi-layer coil countercurrent chromatography (MLCCC) it was finally possible to isolate two different glucoconjugates of the acetylenic diol **21** from Riesling wine (Figure 9) (*23*). The most likely genuine precursor, i.e. glycoconjugated allenic triol **19**, however, has so far not been isolated from wine. Only in enzymatic hydrolyzates of Merlot grapes a detection of the allenic triol **19** as an aglycone was successful (*24*). The intact glucoconjugate **9-O-Glc-19** has for the first time been isolated from leaves of *Lycium halimifolium* (*25*) and later also from *Premna subscandens* (*26*) and *Rosa damascena* flowers (*27*).

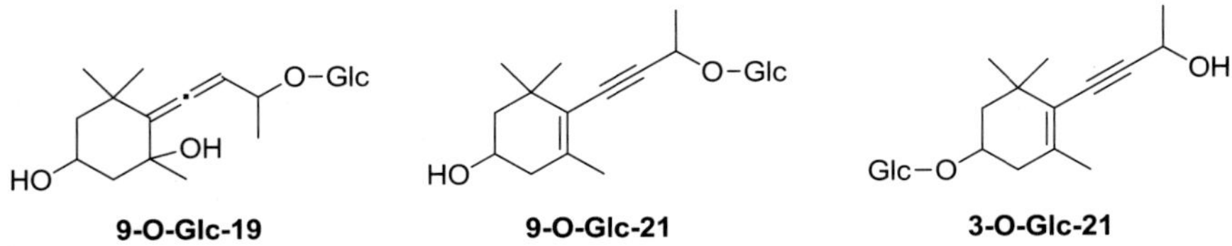


*Figure 9. Glucosidic precursors of β-damascenone **2** (for details cf. text).*

The finding of two different glucoconjugates of acetylenic diol **21** leads us to an important issue and this is the different reactivity of these individual aroma progenitors. In general, it can be said that in case of aglycones with more than one hydroxyl group, the site of glycoconjugation will considerably affect the ratio in which the target volatiles, e. g. β-damascenone **2** and 3-hydroxy-β-damascone **20**, will be formed. In case of the 9-O-glucoconjugate stabilization at carbon atom 9 leads to an enhanced dehydration at position 3 which ultimately will yield higher amounts of the odoriferous ketone **2**. *Vice versa*, stabilization at C-3 allows a Meyer-Schuster rearrangement in the side chain of **9-O-Glc-21** and causes a preferred formation of hydroxyketone **20** (*28*, *29*) (Figure 10).

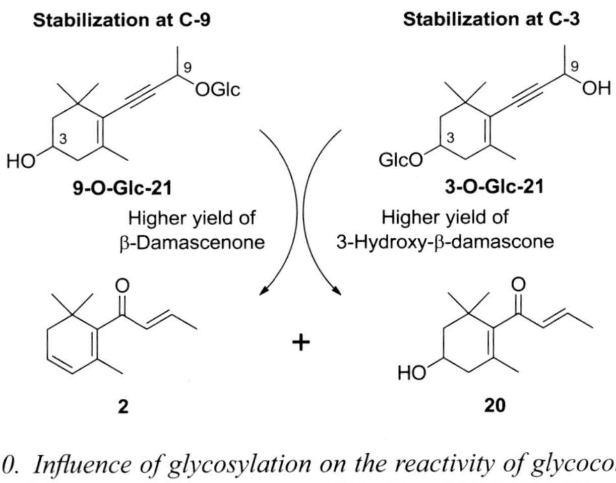


*Figure 10. Influence of glycosylation on the reactivity of glycoconjugated β-damascenone progenitors **9-O-Glc-21/3-O-Glc-21**.*

The existence of different glycoconjugation sites together with the *per se* different relative reactivities of the hydroxyl-groups in the precursor polyols adds to the complexity of the pathways of β-damascenone formation (Figure 11). Whereas the originally suggested pathway (*20*, *21*) via grasshopper ketone **7** and allenic triol **19** mainly yields the stable 3-hydroxy-β-damascone **20** (Pathway A), reaction via acetylenic diol **21** produces a mixture of the target compound **1** and its 3-hydroxyderivative **20** (Pathway B). The recently discovered route (Pathway C) with loss of the C-4 hydrogen gives rise to a formation of megastigma-4,6,7-trien-3,9-diol **24**. After dehydration at C-3 megastigma-3,4-dien-7-yn-9-ol **25** is obtained which almost exclusively (> 90%) will be converted into β-damascenone **2** (*30–34*).

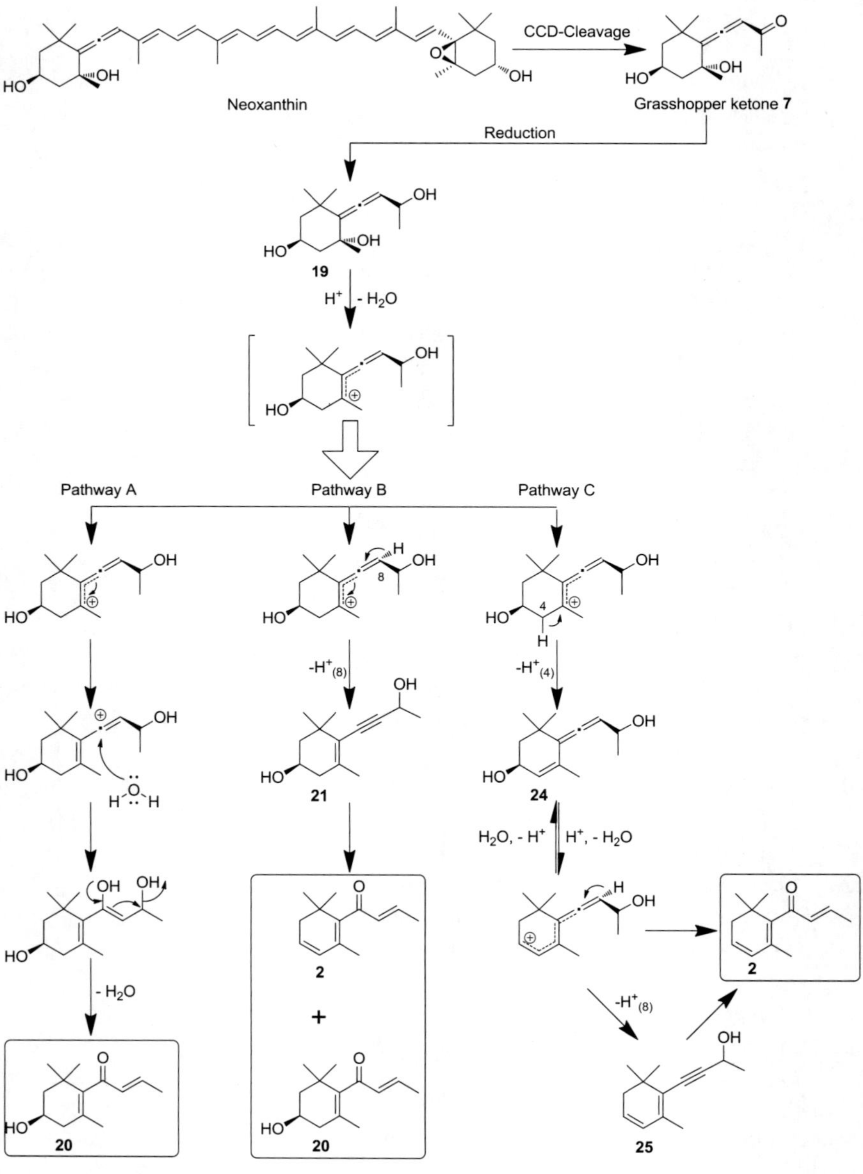


*Figure 11. β-Damascenone **2** formation via acetylenic and allenic intermediates. Adapted with permission from Reference (33) and (34). Copyright 2005and 2008, American Chemical Society.*

## Acknowledgments

Part of this research project was supported by the German Ministry of Economics and Technology (via AiF) and the FEI (Forschungskreis der Ernährungsindustrie e.V., Bonn). Project AiF 16627 N.

PW expresses his thanks to his co-workers who are identified in the references.

## References

1. *Carotenoid-derived Aroma Compounds*; Winterhalter, P., Rouseff, R. L., Eds.; ACS Symposium Series 802; American Chemical Society: Washington, DC, 2002.
2. Baumes, R.; Wirth, J.; Bureau, S.; Günata, Y.; Razungles, A. *Anal. Chim. Acta* **2002**, *458*, 3–14.
3. Winterhalter, P. In *Advances and Challenges in Flavor Chemistry & Biology. Proceedings of the 9th Wartburg Symposium on Flavor Chemistry and Biology*; Hofmann, T., Meyerhof, W., Schieberle, P., Eds.; Deutsche Forschungsanstalt für Lebensmittelchemie: Freising, Germany, 2011; pp 193–200.
4. Williams, P. J.; Sefton, M. A.; Francis, I. L. In *Flavor Precursors - Thermal and Enzymatic Conversions*; Teranishi, R., Takeoka, G. R., Güntert, M., Eds.; ACS Symposium Series 490; American Chemical Society: Washington, DC, 1992; pp 74–86.
5. Simpson, R. F. *Chem. Ind. (London)* **1978**, *1*, 37.
6. Sacks, G. L.; Gates, M. J.; Ferry, F. X.; Lavin, E. H.; Kurtz, A. J.; Acree, T. E. *J. Agric. Food Chem.* **2012**, *60*, 2998–3004.
7. Marais, J.; van Wyk, C. J.; Rapp, A. *S. Afr. J. Enol. Vitic.* **1992**, *13*, 23–32.
8. Gerdes, S. M.; Winterhalter, P.; Ebeler, S. E. In *Carotenoid-derived Aroma Compounds*; Winterhalter, P., Rouseff, R. L., Eds.; ACS Symposium Series 802; American Chemical Society: Washington, DC, 2002; pp 262–272.
9. Karrer, P.; Ochsner, P. *Helv. Chim. Acta* **1948**, *31*, 2093–2097.
10. Stoltz, L. P.; Kemp, T. R.; Smith, W. O., Jr; Smith, W. T., Jr; Chaplin, C. E. *Phytochemistry* **1970**, *9*, 1157–1158.
11. Miginiac, P. *Synth. Commun.* **1990**, *20*, 1853–1856.
12. Eshuis, J. J. W. *Tetrahedron Lett.* **1994**, *35*, 7833–7836.
13. Kotseridis, Y.; Baumes, R.; Skouroumounis, G. K. *J. Chromatogr., A* **1998**, *824*, 71–78.
14. Winterhalter, P. *J. Agric. Food Chem.* **1990**, *38*, 452–455.
15. Güldner, A.; Winterhalter, P. *J. Agric. Food Chem.* **1991**, *39*, 2142–2146.
16. Winterhalter, P. *J. Agric. Food Chem.* **1991**, *39*, 1825–1829.
17. Daniel, M. A.; Capone, D. L.; Sefton, M. A.; Elsey, G. M. *Aust. J. Grape Wine Res.* **2009**, *15*, 93–96.
18. Waldmann, D.; Winterhalter, P. *Vitis* **1992**, *31*, 169–174.
19. Stingl, C.; Knapp, H.; Winterhalter, P. *Nat. Prod. Lett.* **2002**, *16*, 87–93.
20. Isoe, S.; Katsumura, S.; Sakan, T. *Helv. Chim. Acta* **1973**, *56*, 1514–1516.
21. Ohloff, G.; Rautenstrauch, V.; Schulte-Elte, K. H. *Helv. Chim. Acta* **1973**, *56*, 150–31513.

22. Sefton, M. A.; Skouroumounis, G. K.; Elsey, G. M.; Taylor, D. K. *J. Agric. Food Chem.* **2011**, *59*, 9717–9746.
23. Baderschneider, B.; Skouroumounis, G.; Winterhalter, P. *Nat. Prod. Lett.* **1997**, *10*, 111–114.
24. Sefton, M. A. *Aust. J. Grape Wine Res.* **1998**, *4*, 30–38.
25. Näf, R.; Velluz, A.; Thommen, W. *Tetrahedron Lett.* **1990**, *31*, 6521–6522.
26. Sudo, H.; Ide, T.; Otsuka, H.; Hirata, E.; Takushi, A.; Shinzato, T.; Takeda, Y. *Chem. Pharm. Bull. (Tokyo)* **2000**, *48*, 542–546.
27. Suzuki, M.; Matsumoto, S.; Mizoguchi, M.; Hirata, S.; Takagi, K.; Hashimoto, I.; Yamano, Y.; Ito, M.; Fleischmann, P.; Winterhalter, P.; Morita, T.; Watanabe, N. *Biosci., Biotechnol., Biochem.* **2002**, *66*, 2692–2697.
28. Skouroumounis, G. K.; Massy-Westropp, R. A.; Sefton, M. A.; Williams, P. J. In *Progress in Flavour Precursor Studies – Analysis, Generation, Biotechnology*; Schreier, P., Winterhalter, P., Eds.; Allured Publ.: Carol Stream, IL, 1993; pp 275–278.
29. Winterhalter, P.; Skouroumounis, G. In *Biotechnology of Aroma Compounds (Advances in Biochemical Engineering/Biotechnology)*; Berger, R. G., Ed.; Springer Verlag: Heidelberg, 1997; Vol. 55, pp 73–105.
30. Skouroumounis, G. K.; Massy-Westropp, R. A.; Sefton, M. A.; Williams, P. J. *Tetrahedron Lett.* **1992**, *33*, 3533–3536.
31. Skouroumounis, G. K.; Sefton, M. A. *J. Agric. Food Chem.* **2000**, *48*, 2033–2039.
32. Puglisi, C. J.; Elsey, G. M.; Prager, R. H.; Skouroumounis, G. K.; Sefton, M. A. *Tetrahedron Lett.* **2001**, *42*, 6937–6939.
33. Puglisi, C. J.; Daniel, M. A.; Capone, D. L.; Elsey, G. M.; Prager, R. H.; Sefton, M. A. *J. Agric. Food Chem.* **2005**, *53*, 4895–4900.
34. Daniel, M.; Puglisi, C.; Capone, D.; Elsey, G.; Sefton, M. *J. Agric. Food Chem.* **2008**, *56*, 9183–9189.
35. Lee, S.; Seo, M.; Riu, M.; Cotta, J. P.; Block, D. E.; Dokoozlian, N. K.; Ebeler, S. E. *Am. J. Enol. Vitic.* **2007**, *58*, 291–301.
36. Young, P. R.; Lashbrooke, J. G.; Jacobson, D.; Moser, C.; Velasco, R.; Vivier, M. A. *BMC Genomics* **2012**, *13*, 243.

# Chapter 12

# Monitoring Carotenoids and Derived Compounds in Grapes and Port Wines: Impact on Quality

**P. Guedes de Pinho,[1] R. C. Martins,[2] M. A. Vivier,[3] P. R. Young,[3] C. M. Oliveira,[4,5] and A. C. Silva Ferreira*,[3,5]**

**[1]REQUIMTE - Toxicological Laboratory, Biological Science Department, Faculty of Pharmacy, University of Porto, Rua Jorge Viterbo Ferreira, 228, 4050-313 Porto, Portugal**
**[2]BioInformatics - Molecular and Environmental Biology Centre, Universidade do Minho, Campus de Gualtar, 4710-057 Braga, Portugal**
**[3]Stellenbosch University, Private Bag X1, Matieland, 7602, Stellenbosch, South Africa**
**[4]Departamento de Química & QOPNA, Universidade de Aveiro, Campus Universitário Santiago, 3810-193 Aveiro, Portugal**
**[5]Escola Superior de Biotecnologia - CBQF, Universidade Católica Portuguesa, Rua Dr. António Bernardino de Almeida, 4200-072 Porto, Portugal**
***E-mail: asferreira@porto.ucp.pt; antoniof@exchange.sun.ac.za.**

Carotenoids were identified in Port wines where they have been shown to be precursors of norisoprenoids that have floral, fruity or sweet aromas. The finding that young Port wines contain significantly more carotenoids and chlorophyll-like compounds than old Port wines indicated that increases in norisoprenoids during ageing could be correlated with decreases of the aforementioned precursors. In subsequent simulated, forced-ageing experiments, lutein was more sensitive to temperature than β-carotene, and aged wines showed higher β-carotene/lutein ratios compared to young Ports. Norisoprenoid compounds that included β-damascenone, β-ionone, 2,2,6-trimethylcyclohexanone (TCH), 1,1,6-trimethyl-1,2-dihydronaphthalene (TDN), and vitispirane were determined in 14 young and 45-year old (aged) Port wines. In a high oxygen environment, as is the case in Port wine, the levels of these compounds decreased while oxygen was being consumed. However, during barrel ageing, TDN, vitispirane and TCH levels increased significantly while β-ionone and β-damascenone levels decreased. During forced-ageing of Port wines, β-ionone and β-cyclocitral levels increased (2.5 times) in β-carotene-supplemented wines and model wines. Along with these compounds, levels of β-damascenone also increased in both lutein-supplemented wines and model wines.

The role of carotenoids as biomarkers for Port wine quality lead to the development of non-invasive, non-destructive *in vivo* carotenoid quantitation strategies in grapes using vis-SWNIR (visible-short wave near infrared) spectroscopy techniques together with Process Analytical Technologies (PAT) methodologies to provide winemakers with state-of-the-art metabolic images of vineyards for precision winemaking.

## Introduction

In recent years there has been much interest in the contribution of carotenoid-derived breakdown products, such as the norisoprenoids, to wine aroma as they have pleasant aroma descriptors and very low olfactory perception thresholds.

Carotenoids are a heterogeneous group of plant isoprenoids primarily involved in photosynthesis. The norisoprenoids can be formed by direct degradation of carotenoids via chemical degradation (thermal degradation, photo-degradation and (auto)oxidation) or via enzymatic breakdown by several enzymes, such as carotenoid cleavage oxygenases (CCO) also referred to as carotenoid cleavage dioxygenases (CCD), lipoxygenases (LOX), xanthine oxidase, phenoloxidases and peroxidases, resulting in primary non-aromatic cleavage products or directly flavor-active volatiles.

Primary cleavage products can then form a variety of metabolites that are stored as glycoconjugates after an enzymatic glycosylation. These glycoconjugates can then release their volatile aglycone during fermentation via enzymatic processes or acid hydrolysis processes. Norisoprenoids such as TCH (2,2,6-trimethylcyclohexanone), β-damascenone, α- and β-ionone, vitispirane, actinidiol, TDN (1,1,6-trimethyl-1,2-dihydronaphthalene), riesling acetal and TPB (4-(2,3,6-trimethylphenyl)buta-1,3-diene) are responsible for the typical varietal aroma of some grape cultivars (*1–4*).

In this chapter we intend to review the role of carotenoids in grapes in general and in particular in Port wines, as well as the enzymatic and chemical pathways involved in the formation of norisoprenoids, addressing critical questions regarding these important biomarkers for Port wine quality.

## Carotenoids in Grapes

In grape berries the presence of carotenoids is well recognized. In mature grapes, carotenoid concentrations range between 0.8 and 2.5 mg/kg (*5–7*), and the major carotenoids are typically β-carotene and lutein (80-85% of the total carotenoid content) (*7*). The remaining fraction is represented by other minor xanthophylls at µg/kg levels which include neochrome, neoxanthin, violaxanthin, luteoxanthin, flavoxanthin, lutein-5,6-epoxide, zeaxanthin, and *cis* isomers of lutein and β-carotene. In recent years special attention has been given to grape berry carotenoids since they have been identified as precursors of potent aroma compounds in wine, specifically the $C_{13}$-norisoprenoids (*9*, *10*).

Cultivar, climatic conditions, viticultural region, plant water status, exposure to sunlight and ripening stage have all been shown to affect carotenoid concentrations in grape berries (*11–13*). Moreover, the combination of these multiple environmental and viticultural factors may affect the quality of the grapes and therefore the quality of wine.

The carotenoids that have been proposed to be directly involved in the aroma formation of wine are β-carotene and neoxanthin, although lutein and violaxanthin can also be considered because they also undergo breakdown reactions that may produce norisoprenoid compounds (*14*, *15*). This is apparently related to chemical and enzymatic degradation, or it may be due to bioconversion of these compounds.

It is interesting to note that the enzymes known to cleave carotenoids (i.e. CCDs) are typically promiscuous since they have been shown to cleave a variety of different carotenoid- and apocarotenoid substrates *in vitro*. Based on work done in mycorhizzal roots it is currently speculated that the CCD enzymes perform a "recycling" function by degrading the $C_{40}$ carotenoids in the chloroplast and the $C_{27}$ apocarotenoids in the cytosol to form the volatile $C_{13}$ norisoprenoids of interest in wine. The CCD4 (and possibly CCD7) orthologues localized in the chloroplast are hypothesized to be primarily responsible for the cleavage of the $C_{40}$ carotenoids (also localized in the chloroplast) to a $C_{13}$ volatile norisoprenoids and a $C_{27}$ molecule (*16*, *17*). The cytosolically localized CCD1 orthologues are subsequently required for cleaving the $C_{27}$ molecule to volatile $C_{13}$ norisprenoids and a $C_{14}$ molecule. The specific carotenoids targeted for recycling will define the specific volatile norisoprenoid formed (*16*, *18*).

### Biosynthesis of Carotenoids and Its Regulation

The biosynthesis of carotenoids (Figure 1) follows the non-mevalonate pathway in the chloroplasts (*19*) via isopentyl diphosphate (IPP) as a precursor, obtained by condensation of pyruvate and glyceraldehyde-3-phosphate via 1-deoxy-D-xylulose-5-phosphate (*20*, *21*). Britton et al. (*19*) reported that biogenesis of carotenoids takes place in the chloroplast and is an integral part of the chloroplast development.

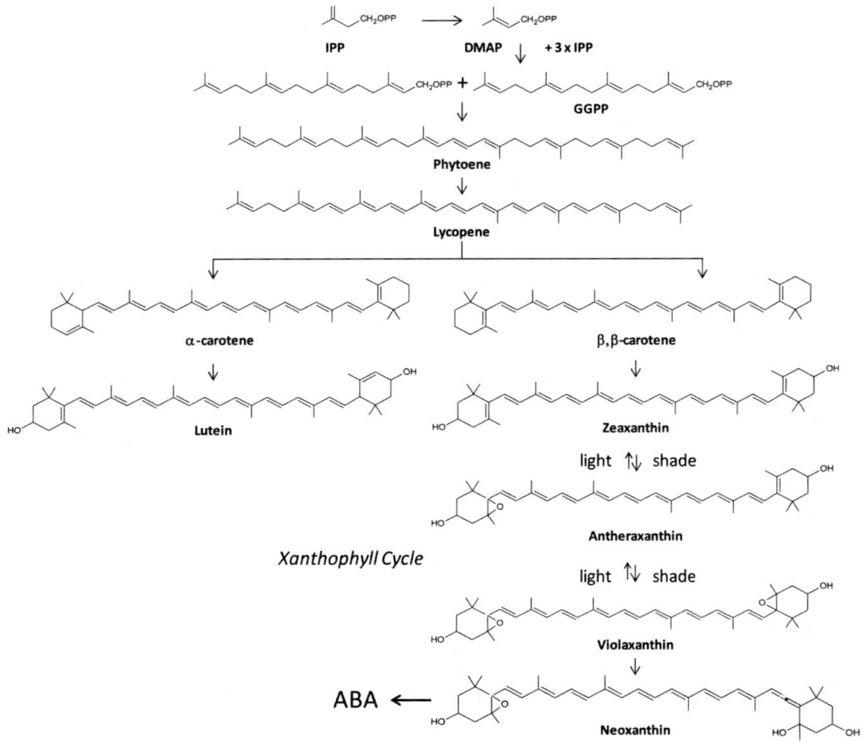


*Figure 1. A simplified diagram of the carotenoid metabolic pathway in plants. ABA - abscisic acid; DMAP - dimethylallyl diphosphate; GGPP - geranylgeranyl diphosphate.*

Carotenoids in unripe grape berries function as light-harvesters (together with chlorophyll) and quenchers of excess light in the photosynthetic systems of the chloroplast (*22, 23*). Carotenoids are associated with multi-protein complexes of plant chloroplast membranes which makes up the photosynthetic systems (PS I and PS II). In these complexes the two main functions of carotenoids in photosynthesis are photo-protection and light harvesting.

The pattern of chloroplast carotenoids in photosynthetically active tissue has been found to be conserved and contains four basic carotenoid-pigments, namely one carotene (β-carotene) and three xanthophylls (lutein, violaxanthin and neoxanthin). Additional minor pigments such as α-carotene, α- and β-cryptoxanthin, isolutein (lutein 5,6-epoxide), zeaxanthin and antheraxanthin (zeaxanthin 5,6-epoxide) were also found. The relative levels of carotenoids in chloroplasts are as follows: lutein 40-57%; β-carotene 25-40%; violaxanthin 9-20%; and neoxanthin 5-15% (*24*).

The carotenoid profile of grapes are cultivar dependent with the skins containing approximately 65% of the total berry carotenoids (predominantly lutein, monoesters of xanthophylls and β-carotene) while the contribution of the pulp is only 35% (*25*). Moreover, in grapes, the content of neoxanthin was found to be three times higher in skin than in pulp, as was the proportion of β-carotene to total carotenoids in these tissues (*25*). Razungles et al. (*6*) suggested that carotenoid concentrations are higher in the skin, since photosynthetic activity is higher in skins.

Recent studies (*26*) have revealed the complex nature of carotenoid regulation at multiple levels. Although relatively well studied biochemically, the transcriptional regulation of this pathway is still not well understood. In this study multiple isoforms were shown to exist for the carotenoid biosynthetic and catabolic genes and enzymes, including the *V. vinifera CCDs (VvCCDs)*. Although the isoforms share a high degree of sequence similarity; they typically have divergent expression patterns. For *VvCCD1.1,VvCCD1.2* (a tandem duplication in grapevine) and *VvCCD4a* the expression profiles increased until véraison, whereas *VvCCD4b* expression increased dramatically during berry ripening. A concomitant decrease in total carotenoid content is also observed. From a plant physiological perspective, the interest in carotenoid cleavage is not limited to their role in the formation of volatile flavor/aroma compounds; since enzymatic cleavage of carotenoids is also involved in the formation of apocarotenoids that function as phytohormones in plants, namely strigolactone and abscisic acid (ABA).

In general, it is becoming increasingly clear that an integrated analysis at the molecular level is required to elucidate function and gene-to-metabolite and/or metabolite-to-metabolite interactions. By using a pathway analysis approach, it was possible to identify a number of the carotenoid metabolic genes with similar profiles that are possibly co-regulated/co-responding (*26*).

## Parameters Affecting Carotenoids Content: Ripening Stage; Viticultural and Climatic Conditions

As discussed in the preceding section carotenoids are generally synthesized from the first stage of grape formation until véraison and then degraded from véraison (during the ripening and maturation stages) (reported by several authors: (*7, 11, 13, 26–30*)) , although these decreases are less prominent during maturation (*5–7*). A study with Portuguese varieties showed that between véraison and maturity, the largest decrease was observed for lutein, from 66% (*13*) to 83% (*12*). In this study, β-carotene gradually decreased reaching a constant level until the harvest date. Similarly, neoxanthin, seemed to decrease during this phase, albeit marginally (*13*).

In more recent works published by Lashbrooke et al. (*30*) and Young et al. (*26*) the photosynthetic pigments (chlorophylls and carotenoids) were monitored at three sampling time-points (green, véraison and harvest). The total carotenoid content decreases approximately 3-fold during ripening/maturation. According to these authors this trend can be mainly attributed to decreases in lutein and β-carotene that represent the most abundant carotenoids in berries (55-60% of the total). Interestingly, the xanthophylls zeaxanthin and antheraxanthin showed an inverse trend by increasing, with zeaxanthin peaking at harvest and antheraxanthin peaking at véraison. The xanthophylls are known to be involved in photoprotection via the violaxanthin and lutein cycles.

The effect of grape variety on carotenoid concentrations in grapes has similarly been studied. Carotenoid profiles were evaluated in eight representative grape varieties from the Douro region of Portugal, over a three year period. The grape varieties with higher carotenoid concentrations were Touriga Fêmea, Tinta Amarela, and Tinta Barroca (*13*, *29*). More recently, carotenoid contents were also evaluated for Merlot (*8*) and four representative wine grape varieties of the Apulian region of Italy (Chardonnay, Merlot, Negroamaro, and Primitivo) (*31*).

The effect of different climatic conditions and/or sunlight exposure on carotenoid concentrations in grapes has been extensively studied (*11*, *27*, *28*, *32*). In general, the highest carotenoid levels occurred in grapes produced in hot regions. On the other hand, it has been shown that variation in light incidence on grapes may affect the berry carotenoid content when comparing sun-exposed and shaded grape bunches (*6*, *13*, *27*, *32*, *33*). During maturity, grapes exposed to sunlight show a significant decrease in carotenoids compared to grapes under shade conditions (*7*, *10*, *11*, *27*, *28*), but it seems that light promotes the increase of specific carotenoids in unripe grapes, i.e. before véraison, compared to the shaded grapes. Due to the role of carotenoids and apocarotenoids to abiotic stress (e.g. the role of the xanthophylls in light stress; and ABA in water stress) their formation is greatly affected by external conditions related to climatic conditions and terroir.

The effect of altitude (terroir) and different cultivars (Tinta Amarela, Tinta Barroca, Souzão, Touriga Franca, Touriga Nacional, Tinta Roriz, Tinto Cão, Touriga Fêmea) on the carotenoid content of grape berries in the Douro Valley of Portugal have been studied by Oliveira et al. (*13*). These authors found that high-elevation terraces, which typically have lower temperatures and higher humidity, during the maturation period, produce grapes with higher carotenoid content. Furthermore, grapes grown with lower vegetative height had higher weight and sugar concentrations. This could be explained by the effect of canopy density and the consequent exposure of bunches to sunlight. For example, grapes grown to lower vegetative heights are less protected from sunlight exposure than grapes grown to higher vegetative height and consequently have lower carotenoid concentrations (*13*).

The effect of plant water status on carotenoid profile was quantitatively determined in Touriga Nacional grape variety from the Douro region of Portugal. The comparison between irrigated (I) and non-irrigated (NI) grapes was followed from 2 weeks before véraison until the ripe stage (*12*). Depending on the soil water retention capacity, the carotenoid contents can be either similar or lower with irrigation treatments (*12*). For soils with high water retention capacity, the total carotenoid content was similar for both NI and I treatments. However, with regard to soils with low water retention capacity, in the I treatment, levels of carotenoids were approximately 60% lower than those found for the NI treatment. Moreover, it seems to be possible to produce higher weight berries (with higher sugar levels) with similar carotenoid contents. On the other hand, soil characteristics had a larger influence than irrigation on the concentration of carotenoids in grapes, resulting in an important viticultural parameter to take into account in aroma precursor formation.

## Composition Profile: A Metabolomic Approach to Plant Water Status

The physiological response of plants to external disturbances is a complex process. In this way, this "multivariate" response constitutes a barrier for the interpretation of plant external interventions. Using bioinformatical signal processing strategies a holistic view of the metabolic phenomenon can be obtained highlighting simultaneous biochemical processes occurring in the cell. Thus, bioinformatics have been developed and applied to data collection of metabolic information to achieve knowledge in plant physiology.

A metabolomic approach was conducted with the variety *V. vinifera* L. cv. Touriga Nacional of the Douro Superior subregion of northern Portugal. Samples were defined in terms of four irrigation regimes (100% of potential evapotranspiration ($ET_0$); 75% of $ET_0$; 50% of $ET_0$; and not irrigated (NI)) and harvested at four maturation times, from véraison to harvest. The extraction and chromatography methodologies were described in Guedes de Pinho et al. (*25*).

In this study, the overall procedure was not targeted to specific compounds, involving rather a set of metabolites, extracted from the grapes collected at the different stages of maturation. Data was obtained from the four different irrigation regimes and focused on the widest possible metabolic profile, justifying the designation "metabolomics: a non-targeted approach".

The Diode-Array Detector (DAD) used in the identification and quantification of carotenoids belongs to a particular group of detectors called "Rich Information Detector " (RIDs),. Each chromatogram can be considered a set of compounds ("scans") obtained by a constant frequency over the time of elution. Each scan contains an absorbance spectrum in a range between 270 and 600 nanometers (nm). Each sample can be represented by a vector (i.e. chromatogram) obtained in a particular wavelength, or as an array surface, where the x-axis represents the retention time; the y-axis represents the absorbance; and the z-axis represents the wavelength range. Using appropriate algebraic algorithms it is possible to calculate a single array resulting from the compression of *n* matrices (chromatographic run) held.

The procedure goes through the initial matrix algebraic conversion dimensions [number of "scans"; wavelengths] on a dimension vector [1; number of "scans" x wavelengths]. The converted vectors can then be grouped by sets of samples with again an array of dimensions [number of samples; number of "scans" x wavelengths]. The treatment of this array signal processing techniques supervised Partial Regression ("Partial Least Squares Regression), or not-supervised by decomposition of Natural Values ("Principal Components Analysis") will lead to a reduction of dimensions being similar to the original array.

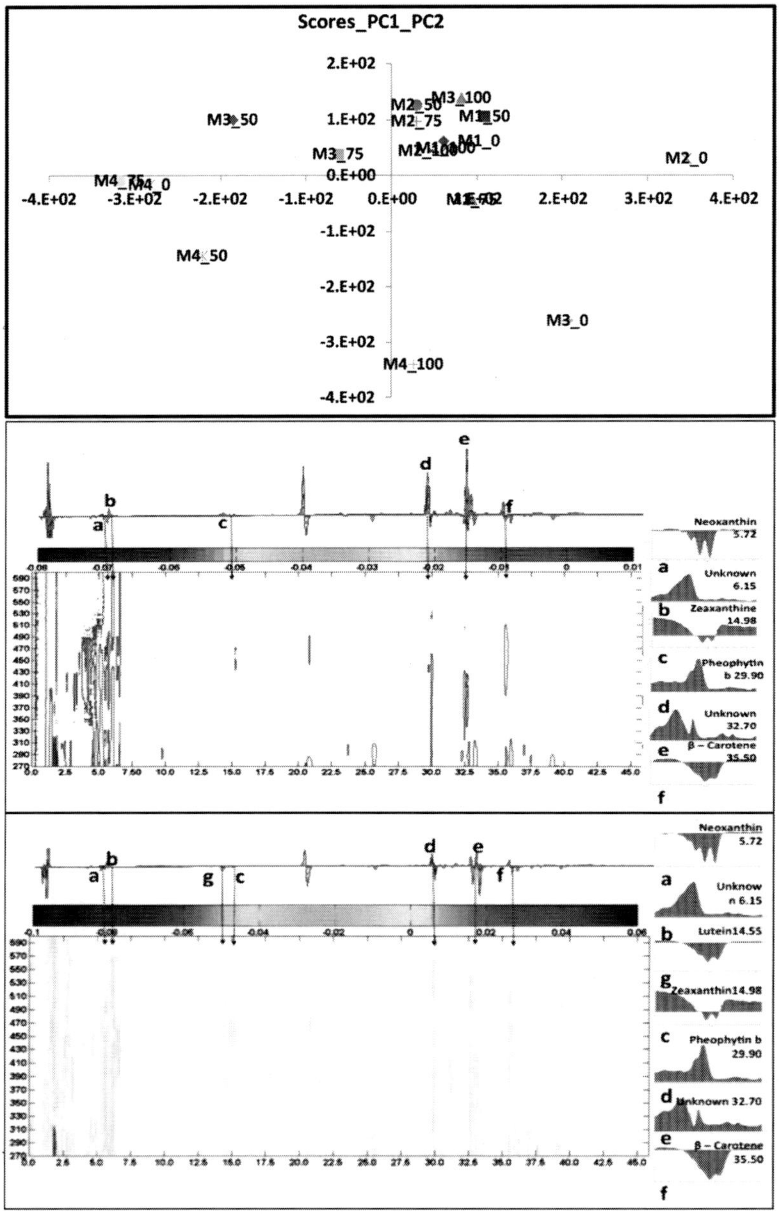


*Figure 2. Principal Components Analysis of the carotenoid contents of grapes obtained with different applied irrigation regimes (100% of $ET_0$ - 100; 75% of $ET_0$ - 75; 50% of $ET_0$ - 50; and not irrigated (0% of $ET_0$ - 0) during the maturation process (M1, M2, M3 and M4): Factor score plot 1-2. Components 1 and 2 account for 82% of the total variance. Loadings of plot 1 (PC1) and Loadings of plot 2 (PC2).*

### PCA Classification

The PCA algorithm is a blind and non-supervised method by which samples are grouped, and relevant features that discriminate between the samples in the time course, can be captured. Time-course metabolic direction can be observed in relevant PC's scores and interpreted by using the relevant PC's loadings. These indicate the wavelength and the compounds that are formed and consumed during the maturation process, as well as their contribution in the different applied irrigation regimes. Chromatograms were normalized by the internal standard (β-apo-caroten-8-al) by mean scaling and division by standard deviation for the assignment of the same degree of importance to the preprocessing resulting metabolites.

In Figure 2, the global metabolic differences are separated in different stages of maturation resulting from different regimes of irrigation. Selecting the 447 nm wavelength, used for the acquisition of carotenoids, it can be seen that there are several compounds that absorb at this wavelength, and are positively (+) or negatively (-) correlated with maturation. The relevant metabolic information filtering, applied to different wavelengths, allows the exploration of other wavelengths, in addition to 447 nm (carotenoids), providing an indispensable aid in support of knowledge generation in plant physiology (Figure 2).

From Figure 2, it can be presumed that one group can be shaped on the positive PC1 plot, accounting for the first stages of maturation M1 and M2. In these stages, carotenoid content is the highest, which was previously noted by other authors. Concerning the plant water status, and in general, the non-irrigated treatment (0% of $ET_0$ - 0) contributed to the highest carotenoid concentration, and it can be seen that M1_0, M2_0, and M3_0 are grouped on the positive PC1 plot. Conversely, the negative PC1 plot accounts for the last stages of maturation (M3 and M4). The loadings interpretation of PC1 (Figure 2) allows the identification of neoxanthin, zeaxanthin and β-carotene, in the negative PC1 plot (last stages of maturation), and pheophytin b and an unknown compound (e), in the positive PC1 plot. These last two compounds are found in the early stages of maturation and can be assumed as maturation markers. Pheophytin b and the compound e, constantly decreased during maturation and are independent of external conditions, e.g. water stress; in contrast neoxanthin, zeaxanthin and β-carotene increased. Similarly, the loading interpretation of PC2 (Figure 2) allows the identification of neoxanthin, lutein, zeaxanthin and β-carotene, in the negative PC2 plot, and pheophytin b and compound e, in the positive PC2 plot. Lutein is identified only in the negative PC2 plot (Figure 2), indicating that this compound has a different behavior from the maturation markers (i.e. pheophytin b and compound e) and is appended to a different metabolic pathway.

## Carotenoid Degradation

The negative correlations observed between the levels of $C_{13}$-norisoprenoids and carotenoids during the grape berry development (*9*) and the transfer of $^{13}C$ markers from carotenoids to $C_{13}$-norisoprenoids in berries (*10*) emphasizes the formation of $C_{13}$-norisoprenoids through carotenoid degradation.

Depending on the carotenoid precursor and the breakdown position, a great variety of carotenoid-based flavour compounds can be generated. This breakdown can take place *via* either non-enzymatic (chemical) or enzymatic routes. In all cases, oxidative breakdown reactions, often but not essentially enzymatic, are the first step (*34*). The non-enzymatic degradation reactions (thermal degradation, photo-degradation and auto-oxidation) are unspecific and result in a random mixture of products, depending on the precursor carotenoid and the reaction conditions (*35–38*). Similarly, co-oxidation by lipoxygenases (LOX), xanthine oxidase, phenoloxidases or peroxidases of carotenoids leads to the unspecific formation of carotenoid breakdown products (*37*). Finally, a family of enzymes (CCDs) specifically degrade carotenoids to C9 or C13 norisoprenoids by cleaving the C(7,8) or the C(9,10) double bond of the conjugated carotenoid chain. These enzymes are all related to the same highly homologous carotenoid cleaving dioxygenase gene family (CCD1, -4, -7 and -8) (*34*).

In later stages of berry development the products of the enzymatic cleavage of carotenoids encoded by the *VvCCDs* (CCDs genes) are known to be potent flavour and aroma compounds (*39*). Interestingly the expression profiles of *VvCCD1.1* and *VvCCD1.2* increase until véraison, and *VvCCD4a* and *VvCCD4b* increase dramatically throughout berry development and total carotenoid content concomitantly decreases (*26*). Previous work performed in berries of Muscat of Alexandria and Shiraz (*40*), Trincadeira (*41*), and Chardonnay and Cabernet Sauvignon (*42*) grapevine cultivars showed that expression of the grapevine *VvCCD1* was induced approaching véraison. VvCCD1 was shown to be functional and can cleave both of lutein and zeaxanthin *in vitro*, but not β-carotene, to form 3-hydroxy-β-ionone (*40*). Evidences of norisoprenoids originated from the hydrolysis of glycoside molecules have previously been reported (*43–45*).

## Relationship between Carotenoids in Grapes and Norisoprenoids in Wines

The relationship between carotenoids in grapes and norisoprenoids in wines has been studied by Oliveira et al. (*29*). In this study, wines produced from grapes with high carotenoid contents (measured at harvest date) have the lowest concentrations of free norisoprenoid compounds (*29*). In contrast, grape varieties with low contents in carotenoid molecules (Touriga Nacional, Souzão, and Tinto Cão) corresponded to wines with the highest levels in volatile norisoprenoidss, namely β-ionone, TDN, and vitispirane. Touriga Nacional wines also have higher contents in terpenols and β-ionone, which in part explains the floral and violet aroma characteristic of these wines. Souzão wines have the highest concentration in TDN and vitispirane.

According to Crupi et al. (*46*) the measure of ΔC (carotenoid content in μg/kg), calculated from the difference of total carotenoid concentration between véraison and maturity could be a useful tool to determine the aroma potential of grapes as it indicates that grapes with higher ΔC have the potential to give wines with higher $C_{13}$-norisoprenoid contents.

## Carotenoids and Norisoprenoids in Port Wine

Carotenoids can be present in very low levels in Port wines (*25*). Knowing that they are precursors of norisoprenoids it was hypothesized that they could be a source of volatiles in Port wines having long periods of ageing.

The reason that carotenoid molecules persist in Port wines and not in table wines is certainly due to the winemaking process. Port wine is a naturally sweet wine produced by interrupting alcoholic fermentation by addition of brandy at the 2nd or 3rd day of alcoholic fermentation; some grape-derived compounds can remain intact in the respective wines. Moreover, the addition of brandy (up to 20% v/v ethanol) may facilitate the solubilization of lipophilic molecules (e.g. the carotenoids and chlorophylls).

The finding that young Port wines contain significantly more carotenoids and chlorophyll-like compounds than old Port wines (*47*) indicated that an increase in norisoprenoids during ageing could be correlated with a decrease of the aforementioned precursors. This was subsequently proven in simulated experiments, monitoring the changes in the β-carotene/lutein concentration ratio (*47*, *48*). In these studies, forced-aging experiments indicated that lutein is more sensitive to temperature than β-carotene, and that aged wines showed higher ratios of β-carotene/lutein concentrations compared to new Ports. Moreover, the last study (*48*) has confirmed that levels of β-ionone and β-cyclocitral increased (2.5 times) in both wine and a wine model solution supplemented with β-carotene (*48*). Along with these compounds, the same behavior was observed for β-damascenone in a wine and a wine model solution supplemented with lutein (Silva Ferreira et al., 2008). In addition, two unknown carotenoid degradation products were identified by HPLC/DAD (reverse phase λmax: 422; 445; 475 and 422; 445; 472) in a forced-aged wine supplemented with lutein (*48*).

Norisoprenoid compounds, such as β-damascenone, β-ionone, 2,2,6-trimethylcyclohexanone (TCH), 1,1,6-trimethyl-1,2-dihydronaphthalene (TDN) and vitispirane were determined in 14 young Port wines and 45 old Port wines and, an experimental protocol was performed in order to determine which technological parameter (dissolved $O_2$, free $SO_2$ levels, pH and time/temperature) was related with the formation/consumption of these molecules. These five norisoprenoid molecules were equally affected by the selected parameters and a similar profile with time was observed. The synergistic effects of increasing temperature and lowering pH had the largest impact. For samples treated with high oxygen regimes (saturation), the levels of all considered norisoprenoids decreased after a certain concentration of oxygen consumed (10 mg/L). Nevertheless, during Port wine ageing in barrels, corresponding to the 45 year old Port wines, two different behaviours can be observed: TDN, vitispirane and TCH increased significantly whilst a decrease of levels of β-ionone and β-damascenone with Port barrel ageing was observed. It was also calculated that "over 40 year" old Port wines have, 15, 5 and 3 times higher levels of TDN, vitispirane and TCH, respectively, than the young Ports (*49*).

Nineteen compounds with carotenoid or chlorophyll-like structures were present in Port wines, where the young wines showed higher total carotenoid content and chlorophyll-like compounds compared to aged Ports, with lutein and β-carotene as major carotenoids (*47*). Additionally, aged wines showed higher ratios of β-carotene/lutein concentration compared to new Ports (*47*). The carotenoid degradation and the formation of volatiles were determined recurring to a forced ageing simulation of a Port wine supplemented with lutein or β- carotene (*48*). Results showed that the percentage of decrease of lutein and β-carotene was, respectively, 95% and 10%, indicating that lutein was more sensitive to degradation than β-carotene. Conversely, β-ionone and β-cyclocitral where identified as the major aroma molecules released in a Port wine supplemented with β-carotene, with increases up to 0.2 and 0.5 μg/L, respectively. Furthermore, β-damascenone has been firstly identified in Port wine supplemented with lutein. Although β-carotene and lutein are both carotenoids; it is interesting to note that β-carotene is a carotene - a pure hydrocarbon molecule consisting of only C-H - whereas lutein is a xanthophyll, or an oxygenated carotenoid. Studies on purified carotenoids have noted differential susceptibilities of carotenoids to degradation (*30*).

## Monitoring Carotenoids in Grapes - Vine PAT

The physiological response of plants to external perturbation is complex and occurs at different levels of their metabolism, therefore high-throughput methodologies are required to extract relevant information. In this context, spectroscopic methodologies present the greatest potential for metabolic studies in biological systems.

Variance imaging is a technique that plots the variance decomposition (singular value decomposition) of the spectral data in each coordinate, similar to a heat map. Such maps allow one to use spectroscopy as a high-throughput system to identify and classify grape quality by spectral feature similarity. As the decomposition uses only the relevant singular values, the user will be able to identify the causes of relevant spectral differences among the different points of the mesh.

Metabolic imaging is a technique that plots the expected composition of the grapes at each point of the mesh by the supervised mathematical modelling of the spectra vs. the chemical composition obtained by chemical analytical methodologies (e.g. HPLC-DAD, LC-MS and GC-MS). Under these circumstances, the system calculates the metabolic composition using the obtained spectra.

The workflow for a vis-SWNIR (200- 1200 nm, fiber optics probes to obtain diffusive reflectance measurement) "Hyperspectral" spectroscopy technique, starts by acquiring the spectra at each point of an image; as a non-destructive, "*in situ*", "*in vivo*" measurement and in real-time to obtain information of the metabolic state of ripeness of the grapes, allowing a detailed view of the composition of the vineyard over time. This new tool will provide more information that can be used to assist the development of new approaches to wine production.

Pre-processing is necessary to correct two of the most important systematic variance from spectral information due to: i) light intensity that reaches the detection; and ii) scattering correction due to the irregular shape of cheese surface. Light intensity is given by the number of photons that reach the detection, and in the linear operational region of the spectrum (Charge Coupled Device - CCD) is proportional to the photon/s at each wavelength, given the quantum efficiencies for each pixel of the CCD detection. Afterwards, possible noise was removed by applying a convolution filter (*50*) with length = 4 and signal order = 2, which removes the high frequency noise from initial spectra. Scattering correction was performed by the Robust Multiplicative Scattering Correction (*51*, *52*), which is a modification of the algorithm for Multiplicative Scattering Correction.

In order to develop a hyperspectral image of the vineyard, the system records the spectra in a pre-established mesh with defined nodal points. The mesh is obtained by Voronoi tessellation, and produces triangular elements were vertices are the nodal points, corresponding to a vine sample. This is thereafter used for signal processing to obtain the variance imaging and metabolic imaging.

### Variance and Metabolic Imaging

Variance and metabolic imaging is provided for each relevant component or substance previously calibrated. The user selects the principal component that must be observed, and the system provides the super-imposed image over the mesh and vineyard (Figure 3). Similar colours in these images provide the information on grapes clusters similarity in each component. By using this information, it is possible to identify possible causes of differences of non-supervised metabolic differences.

Metabolic imaging is provided to the user by selecting the metabolite to view in the user interface. The user can navigate the metabolic composition by clicking at each node.

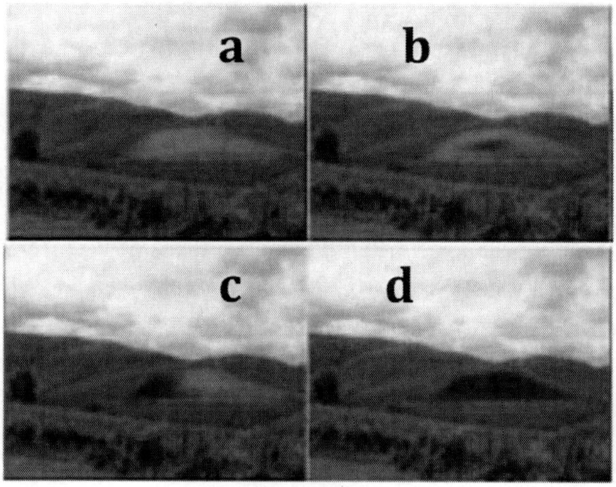


*Figure 3. Time course metabolomic imaging for β-carotene from vérasion (19th of July **a**), (26 th of July **b**), (9th of August **c**) to harvest (30th of August **d**).*

### Time-Course Metabolomics

Figure 3 presents the changes in the β-carotene fraction in a vineyard from 19 July (before vérasion) to 30 August (harvest), showing undetectable levels to maximum concentration (1 $mg.kg^{-1}$).

Results showed that β-carotene content decreased during ripening, which is in agreement with previous works, where the presence of carotenoids in grape berries showed that β-carotene and several xanthophylls are abundant before véraison, and tend to decrease during ripening. Nevertheless, the image shows a high degree of variability of β-carotene along the field during maturation. For example, it is possible to observe that in the 1st sampling point (Figure 3 a; véraison) the highest content of β-carotene is present in the upper-left corner. This shifts to the center in the next frames and attains lower levels at harvest. At harvest it is also observable that lower concentrations exist in the right side.

The examples presented demonstrated the usefulness and advantage of having the Spectroscopy system for the collection of grape metabolic information. The system allows the vineyard manager to monitor the grape maturation process and allows the collection of information about the variance of "grape quality".

The winemakers can also compare the historical records containing the information collected in the different campaign as a way to adjust and plan corrective actions in the vineyard.

## Conclusion

An understanding of the parameters that affect carotenoid biosynthesis in grapes enables the manipulation/modulation of the levels of these precursors of high impact aroma compounds in wine. The interrelationship of carotenoid molecules (β-carotene and lutein) and norisoprenoids plays an important role in Port wine aroma quality. Using *in vivo* data acquisition and a metabolomics data processing pipeline allows real-time monitoring of carotenoids – an ideal tool for management of "grape quality".

## References


1. Sefton, M. A.; Skouroumounis, G. K.; Massy-Westropp, R. A.; Williams, P. *Aust. J. Chem.* **1989**, *42*, 2071–84.
2. Skouroumounis, G. K.; Massy-Westropp, R. A.; Sefton, M. A.; Williams, P. J. *Tetrahedron Lett.* **1992**, *33*, 3533–3536.
3. Freitas, V.; Ramalho, P.; Azevedo, Z.; Macedo, A *J. Agric. Food Chem.* **1999**, *47*, 4327–4331.
4. Janusz, A.; Capone, D. L.; Puglisi, C. J.; Perkins, M. V.; Elsey, G. M.; Sefton, M. A. *J. Agric. Food Chem.* **2003**, *51*, 7759–7763.
5. Razungles, A.; Bayonove, C.; Cordonnier, R.; Baumes, R. *Vitis* **1987**, *26*, 183–191.
6. Razungles, A.; Bayonove, C.; Cordonnier, R.; Sapis, J. *Am. J. Enol. Vitic.* **1988**, *39*, 44–48.

7. Razungles, A.; Babic, I.; Sapis, J.; Bayonove, C. *J. Agric. Food Chem.* **1996**, *44*, 3821–3825.
8. Kamffer, Z.; Bindon, K. A.; Oberholster, A. *J. Agric. Food Chem* **2010**, *58*, 6578–6586.
9. Razungles, A.; Gunata, Z.; Pinatel, S.; Baumes, R. L.; Bayonove, C. L. *Sci. Aliments* **1993**, *13*, 59–72.
10. Baumes, R.; Wirth, J.; Bureau, S.; Gunata, Y.; Razungles, A. *Anal. Chim. Acta* **2002**, *458*, 3–14.
11. Marais, J.; van Wyk, C.; Rapp, A. *S. Afr. J. Enol. Vitic.* **1991**, *12*, 64–69.
12. Oliveira, C.; Silva Ferreira, A. C.; Hogg, T.; Alves, F.; Mendes Pinto, M.; Guedes de Pinho, P. *J. Agric. Food Chem.* **2003**, *51*, 5967–5971.
13. Oliveira, C.; Silva Ferreira, A. C.; Costa, P.; Guerra, J.; Guedes de Pinho, P. *J. Agric. Food Chem.* **2004**, *52*, 4178–4184.
14. Winterhalter, P.; Rouseff, R. Carotenoid-Derived Aroma Compounds: An Introduction. In *Carotenoid-Derived Aroma Compounds*; Winterhalter, P., Rouseff, R., Eds.; ACS Symposium Series 802: American Chemical Society: Washington, DC, 2002; pp 1−17.
15. Wahlberg, I.; Eklund, A. M. Degraded carotenoids. In *Carotenoids: Biosynthesis and Metabolism*; Britton, G., Liaasen-Jensen, S., Pfander, H., Eds.; Birkäuser Verlag: Berlin, 1998; Vol. 3, pp 195−216.
16. Floss, D. S.; Schliemann, W.; Schmidt, J.; Strack, D.; Walter, M. H. *Plant Physiol.* **2008**, *148*, 1267–1282.
17. Floss, D. S.; Walter, M. H. *Plant Signaling Behav.* **2009**, *4*, 172–175.
18. Ilg, A.; Yu, Q. J.; Schaub, P.; Beyer, P.; Al-Babili, S. *Planta* **2010**, *232*, 691–699.
19. Britton, G. *Z. Naturforsch.* **1979**, *34*, 979–985.
20. Lichtenthaler, H. K.; Schwender, J.; Disch, A.; Rohmer, M. *FEBS Lett.* **1997**, *400*, 271–274.
21. Bouvier, F.; Rahier, A.; Camara, B. *Prog. Lipid Res.* **2005**, *44*, 357–429.
22. Van den Berg, H.; Faulks, R.; Fernando Granado, H.; Hirschberg Olmedilla, B.; Sandmann, G.; Southon, S.; Stahl, W. *J. Sci. Food Agric.* **2000**, *80*, 880–912.
23. Krinsky, N. I. *Pure Appl. Chem.* **1979**, *51*, 649–660.
24. Gross, J. In *Pigments in Vegetables: Chlorophylls and Carotenoids*; Van Nostrand Reinhold: New York, 1991; pp 1−351.
25. Guedes de Pinho, P.; Ferreira, A. C.; Mendes Pinto, M.; Gomez Benitez, J.; Hogg, T. *J. Agric. Food Chem.* **2001**, *49*, 5484–5488.
26. Young, P. R.; Lashbrooke, J. G.; Alexandersson, E.; Jacobson, D.; Moser, C.; Velasco, R.; Vivier, M. A. *BMC Genomics* **2012**, *13*, 243.
27. Bureau, S.; Razungles, A.; Baumes, R.; Bayonove, C. *Vitic. Enol. Sci.* **1998**, *53*, 64–71.
28. Razungles, A.; Bureau, S.; Bayonove, C.; Baumes, R. Effect de l'ensoleillement sur les teneurs en precurseurs d'aromes des baies de Syrah. *OIV* **1999**, *11*, 382–388.
29. Oliveira, C.; Barbosa, A.; Silva Ferreira, A. C.; Guerra, J.; Guedes de Pinho, P. *J. Food Sci.* **2006**, *71*, S1–S7.

30. Lashbrooke, J. G.; Young, P.; Strever, A. E.; Stander, C; Vivier, M. A. *Aust. J. Grape Wine Res.* **2010**, *16*, 349–360.
31. Crupi, P.; Coletta, A.; Milella, R. A.; Palmisano, G.; Baiano, A.; La Notte, E.; Antonacci, D. *J. Food Sci.* **2010**, *75*, S191–S198.
32. Bureau, S.; Baumes, R.; Razungles, A. *J. Agric. Food Chem.* **2000**, *48*, 1290–1297.
33. Bindon, K. In *Influence of Partial Root Zone Drying on Aspects of Grape and Wine Quality*; The University of Adelaide: Adelaide, South Australia, 2004; Chapter 9, pp 182−204.
34. Fleischmann, P.; Zorn, H. In *Carotenoids, Vol. 4: Natural Functions*; Britton, G, Liaaen-Jensen, S, Pfander, H, Eds.; Birkhäuser: Basel, 2008; pp 341−366.
35. Kanasawud, P.; Crouzet, J. C. *J. Agric. Food Chem.* **1990**, *38*, 237–243.
36. Crouzet, J.; Kanasawud, P.; Sakho, M. In *Carotenoid-Derived Aroma Compounds*; Winterhalter, P., Rouseff, R., Eds.; ACS Symposium Series 802; American Chemical Society: Washington, DC, 2002; Chapter 9, pp 115-129.
37. Winterhalter, P. In *Flavor Precursors*; Teranishi, R., Takeoka, G. R., Guntert, M., Eds; ACS Symposium Series 490; American Chemical Society: Washington, DC, 1992; pp 98−115.
38. Mordi, C. R.; Walton, J. C.; Burton, G. W.; Hughes, L.; Ingold, K. U.; Lindsay, D. A. *Tetrahedron Lett.* **1991**, *32*, 4203–4206.
39. Walter, M. H.; Floss, D. S.; Strack, D. *Planta* **2010**, *232*, 1–17.
40. Mathieu, S.; Terrier, N.; Proccureur, J.; Bigey, F.; Günata, Z. *J. Exp. Bot.* **2005**, *56* (420), 2721–2731.
41. Fortes, A M.; Agudelo-Romero, P.; Silva, M. S.; Ali, K.; Sousa, L.; Maltese, F.; Choi, Y. H.; Grimplet, J.; Martinez-Zapater, J. M.; Verpoorte, R. *BMC Plant Biol.* **2011**, *1*, 149–183.
42. Deluc, L.; Bogs, J.; Walker, A. R.; Ferrier, T.; Decendit, A.; Merillon, J. M.; Robinson, S. P.; Barrieu, F. *Plant Physiol.* **2008**, *147*, 2041–2053.
43. Gunata, Y. Z.; Bayonove, C. L.; Baumes, R. L.; Cordonnier, R. E. *J. Chromatogr.* **1985**, *331*, 83–90.
44. Di Stefano, R.; Bottero, S.; Pigello, R.; Borsa, D.; Bezzo, G.; Corino, L. *L'Enotecnico* **1998**, *3*, 63–74.
45. Skouroumounis, G. K.; Massy-Westropp, R. A.; Sefton, M. A.; Williams, P. J. *Tetrahedron Lett.* **1992**, *33*, 3533–3536.
46. Crupi, P.; Coletta, A.; Antonacci, D. *J. Agric. Food Chem.* **2010**, *58*, 9647–9656.
47. Mendes Pinto, M. M.; Silva Ferreira, A. C.; Caris-Veyrot, C.; Guedes de Pinho, P. *J. Agric. Food Chem.* **2005**, *53*, 10034–10041.
48. Silva Ferreira, A. C.; Monteiro, J.; Oliveira, C.; Guedes de Pinho, P. *Food Chem.* **2008**, *110*, 83–87.
49. Silva Ferreira, A. C.; Guedes de Pinho, P. *Anal. Chim. Acta* **2004**, *513*, 169–176.
50. Savitzky, A.; Golay, M. J. E. *Anal. Chem.* **1964**, *36* (8), 1627–1639.
51. Gallagher, N. B.; Blake, T. A.; Gassman, P. L. *J. Chemom.* **2005**, *19* (5-7), 271–281.
52. Martens, H.; Nielsen, J. P.; Engelsen, S. B. *Anal. Chem.* **2003**, *75* (3), 394–404.

# Biotechnology and Processing Effects

## Chapter 13

# Biotechnological Production of Norisoprenoid Aroma Compounds

**K. Schmidt, K. Kunkel, R. Szweda, A. Portz, M. A. Fraatz, and H. Zorn***

**Institute of Food Chemistry and Food Biotechnology, Justus Liebig University, Heinrich-Buff-Ring 58, 35392 Giessen, Germany**
***E-mail: holger.zorn@lcb.chemie.uni-giessen.de.**

Norisoprenoid aroma compounds such as $\beta$-ionone and $\beta$-damascenone often represent highly attractive flavor-active organic molecules. In plants, they are formed by an enzyme catalyzed, regio-specific oxidative cleavage of carotenoid precursors. As the natural occurrence of norisoprenoids is limited to trace amounts, it is an alluring idea to copy the plant biosynthetic pathways in biotechnological systems. The best investigated approaches include the co-oxidation of carotenoids with the lipoxygenase/linoleic acid system, the release of bound precursors by means of glycosidases, and the use of recombinant carotenoid cleavage dioxygenases.

In an alternative approach, fungal peroxidases secreted by basidiomycetes have been shown to efficiently degrade various carotenes and xanthophylls, and carbon-13 and carbon-10 norisoprenoids represented the main volatile degradation products in each case. The carotenoid degrading enzymes were purified from submerged cultures and characterized on a molecular level.

## Introduction

Norisoprenoid aroma compounds are formed in fruit and flowers via an enzyme catalyzed regio-selective oxidative cleavage of various precursor carotenoids. Prominent examples include the carbon-13 norisoprenoids $\beta$-ionone and $\beta$-damascenone, which e.g. shape the scents of roses. $\beta$-Ionone and *β*-damascenone have odor thresholds of 0.007 and 0.009 ppm, respectively (*1*), and they thus represent highly flavor-active organic molecules (*2–4*). As the natural occurrence of norisoprenoids in the producer plants is generally limited to trace amounts, solvent extraction is not economically feasible. Alternatively, it is an alluring idea to copy the plant biosynthetic pathways in biotechnological systems. According to American and European legislation, the thus produced norisoprenoids may be labeled as "natural flavors" which represents an enormous competitive advantage.

### Co-oxiation of Carotenoids

One of the best investigated biotechnological approaches for the release of norisoprenoids from carotenoids is the so-called co-oxidation of carotenoids. The carotenoids are oxidized by free-radical species which are generated with the lipoxygenase/linoleic acid or xanthine oxidase/acetaldehyde systems (*5, 6*). The mechanism of co-oxidative carotenoid degradation probably involves enzymatically generated peroxy radicals from which a carotenoid radical and then a peroxycarotenoid radical are produced. In a xanthine oxidase/acetaldehyde co-oxidation system, neoxanthin was degraded to *inter alia β*-damascenone and its proposed precursor grasshopper ketone (*7*). A major drawback of the co-oxidation approach is its low specificity. In systems containing *β*-carotene with either soybean or recombinant pea lipoxygenases, more than 50 products were detected by means of liquid chromatography. Thus, the lipoxygenase initiated oxidation of carotenoids occurs randomly, both at the ring double bonds and at different positions along the conjugated polyene chain (*7*). Therefore, the product yields obtained from the co-oxidative degradation of carotenoids are typically too low for industrial flavor production.

### Release of Bound Precursors by Means of Glycosidases

In plants, norisoprenoid glycosides are often formed and stored until flavors are released by enzymatic hydrolysis of the glycosides at maturation or ripening (*8–10*). The release of free norisoprenoids by the action of specific glycosidases is especially relevant for the aroma of juices and wines (*11*). As discussed above for free norisoprenoids, the concentrations of norisoprenoid glycosides in plants are limited to trace amounts, and their enzymatic release does thus not allow for production of industrially relevant flavor amounts.

### Carotenoid Cleavage Dioxygenases (CCDs)

Highly potent carbon-13 norisoprenoids like *β*-ionone are released from tetraterpenoid carotenoids by cleavage of the C(9,10) or C(9′,10′) double bond, respectively, of the carotenoids' polyene chain. *At*CCD7, an enzyme cloned from *Arabidopsis thaliana* catalysed *in vitro* a specific C(9,10)-cleavage of *β*-carotene to the products *β*-ionone and *β*-apo-10′-carotenal. *At*CCD7 contains a potential chloroplast targeting sequence, which is consistent with a role in metabolism of carotenoids that are located in chloroplasts (*12, 13*). When

*At*CCD1, another carotenoid cleaving enzyme of *A. thaliana,* was expressed in *E. coli*, the recombinant enzyme was able to cleave a number of carotenoids, *e.g.* *β*-carotene and lutein, symmetrically at both the C(9,10) and C(9′,10′) positions of the polyene backbone to produce a $C_{14}$ dialdehyde and two $C_{13}$ norisoprenoid products (*14*).

In the meantime, various enzymes with properties similar to those of *At*CCD1 have been characterized *e.g.* from roses (*15*), tomato, crocus, petunia, rice, and maize (reviewed in (*16*)). An appropriate substrate delivery systems such as micelles or liposomes may help to overcome the current limitations in delivering the non-polar substrates to the enzymatic cleavage reaction (*17*).

## Results and Discussion

In an approach to use side streams of the food industry as growth substrates for edible mushrooms, submerged cultures of various fungi were supplemented with *e.g.* carrot peels as the sole carbon source. Surprisingly, a number of clearly carotenoid derived volatiles were detected in the aroma extracts of some submerged cultures of basidiomycetes (Figure 1).

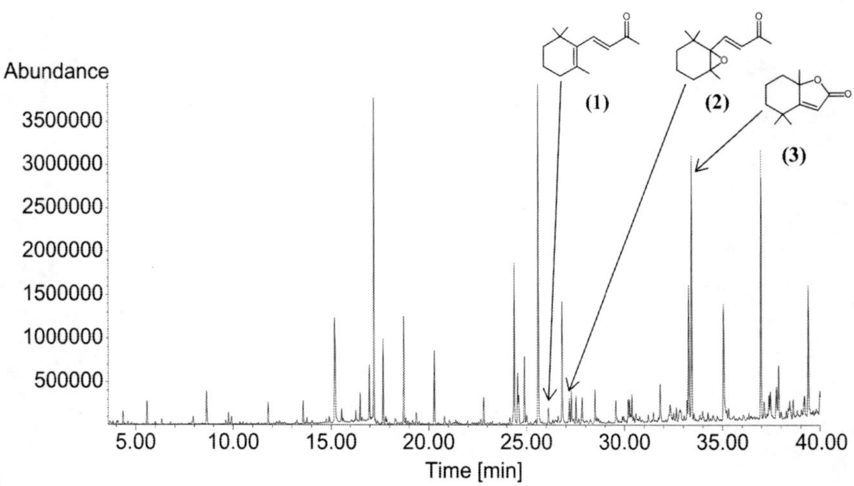


*Figure 1. GC-MS chromatogram of an aroma extract of a submerged culture of the basidiomycete Wolfiporia cocos grown with carrot peels as sole carbon source; **(1)** β-ionone, **(2)** 5,6-epoxy-β-ionone, **(3)** dihydroactinidiolide.*

As these norisoprenoids were not detected in the control experiments with the fungal cultures growing on standard media and media containing the carrot peels as substrate only, a degradation of *β*-carotene by fungal enzymes was most likely.

## Carotenoid Degradation by Fungal Extracellular Peroxidases

To identify potent carotenoid degrading fungi, and thus potential producers of norisoprenoid aroma compounds, more than 50 basidiomycetes and yeasts were screened on *β*-carotene containing agar plates. With ten species, a bleaching halo was observed around the fungal mycelia, indicating a secretion of carotenoid degrading extracellular enzymes into the media. Based on this observation, the positive strains were transferred into submerged cultures. While the basidiomycetous strains degraded emulsified *β*-carotene nearly quantitatively within 12 h, only dihydroactinidiolide was identified as carotenoid degradation product in the respective aroma extracts. Interestingly, when the fungal mycelia were removed prior to the addition of emulsified *β*-carotene to the media, 2-hydroxy-2,6,6-trimethylcyclohexanone, *β*-cyclocitral, and *β*-ionone were formed in addition to dihydroactinidiolide (*18*). As already concluded from the bleaching halos in the surface cultures, extracellular enzymes were obviously responsible for the release of norisoprenoids from *β*-carotene.

To characterize the responsible enzymes, a fast and reliable photometric enzyme assay was developed based on a method reported by (*19*). Therefore, the time-dependent decrease of absorbance of an aqueous *β*-carotene solution was monitored at 450 nm in a thermostatable spectral photometer (Figure 2).

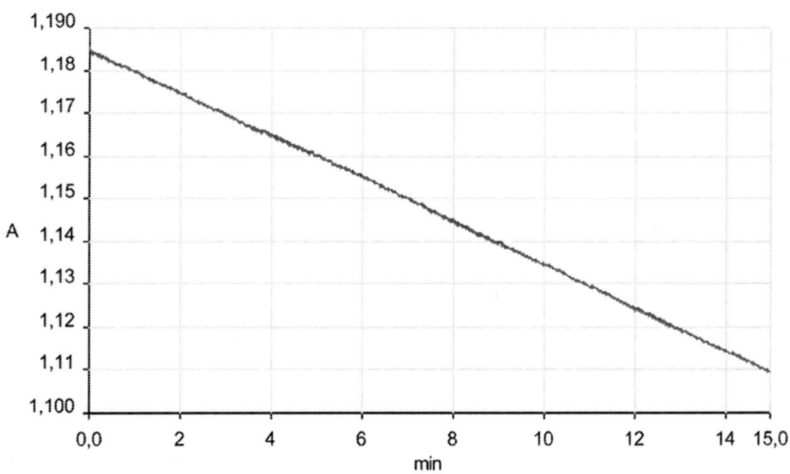


*Figure 2. Time-dependent decrease of absorbance of a β-carotene emulsion as a measure of carotenoid degrading enzyme activity.*

Two of the screened fungi showed in this assay comparatively high extracellular carotenoid degrading enzyme activities, and were thus selected for further investigation.

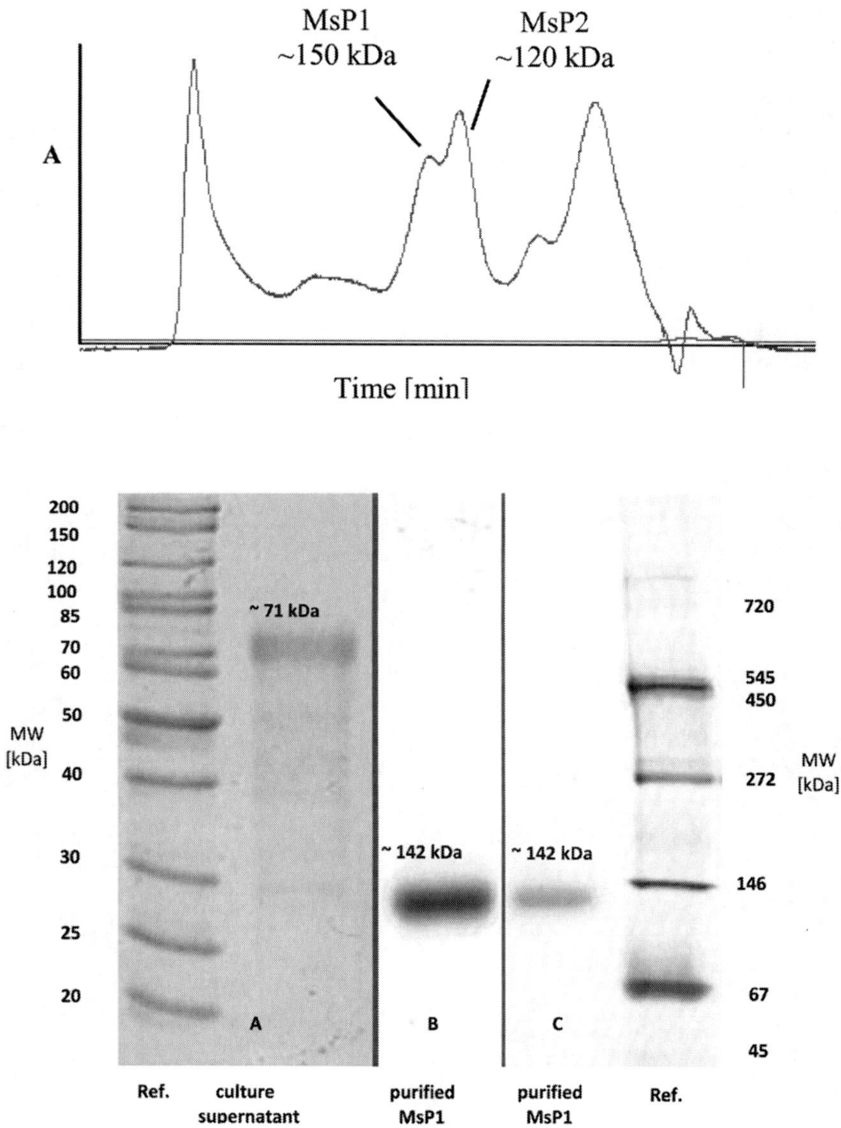


*Figure 3. Size-exclusion chromatography as second chromatographic purification step (top) and SDS-PAGE analysis of MsP1; A: denaturing conditions; B: native gel (SERVAGel™ N Native Gel), activity staining with ABTS; C: native gel, colloidal Coomassie blue staining (bottom).*

From the supernatants of the garlic fungus *Marasmius scorodonius* (synonym *Mycetinis scorodonius)*, two carotenoid cleaving enzymes were purified by means of fast protein liquid chromatography, including the steps ion exchange (IEX) and size exclusion chromatography (SEC). Based on their dependence on hydrogen peroxide as co-factor, they were named as *Marasmius scorodonius* peroxidases 1 and 2 (MsP1 & MsP2). The UV-VIS absorption spectra of purified MsP1 and MsP2 with absorption maxima of 406 and 405 nm (Soret band), respectively, indicated the presence of heme groups. This was further confirmed by semi-native SDS-PAGE electrophoresis using a heme specific staining with 3,3′,5,5′-tetramethylbenzidine (*20*).

Under the non-denaturing conditions of SEC, MsP1 and MsP2 showed molecular masses of ~140 and ~120 kDa, respectively. Likewise, a molecular mass of ~142 kDa was estimated from native PAGE gels for purified MsP1, while a molecular mass of ~70 kDa was detected under denaturing conditions (Figure 3). In good agreement with the electrophoretic data, molecular weights of 64.2 kDa and of 52.4 kDa were deduced for MsP1 and MsP2, respectively, from MALDI-TOF experiments (*20*).

These data strongly suggest a homo-dimeric structure of MsP1. Further support for this hypothesis was obtained from synchrotron small-angle X-ray scattering (SAXS), differential scanning calorimetry (DSC), and temperature dependent fluorescence spectroscopy (*21*). In temperature and pressure dependent activity assays, MsP1 showed activity maxima between 55 and 60 °C and at ~500 bar, respectively.

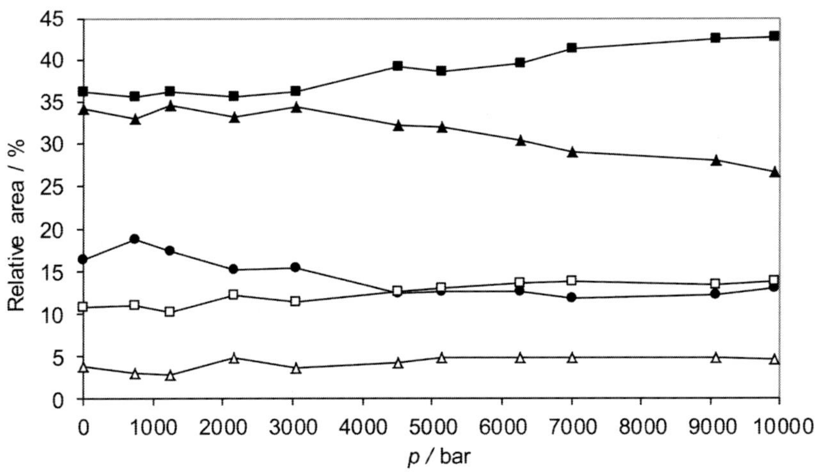


*Figure 4. Pressure dependence of the secondary structure elements of MsP1 as revealed from the FT-IR spectra at 30 °C; the relative peak areas of the subbands of MsP1 are assigned to the following secondary structures: ▪ random coil structure, ▴ α-helices, ● loops and turns, □ intramolecular β-sheets, ▵ intermolecular β-sheets.*

Surprisingly, no significant changes in the enzyme's secondary structure elements were observed under high pressure conditions as revealed from Fourier transformed infrared spectra (FT-IR; Figure 4). With increasing pressure, no apparent change in the band areas were detected until ~3,000 bar, where the

$\alpha$-helical content started to decrease, and the unordered fraction increased. Even above ~3 kbar (at 30 °C), only a minor increase in unordered structures was observed, indicating a high pressure stability of MsP1 (*21*).

By means of electrospray tandem mass spectrometry and by N-terminal Edman degradation, peptide sequences of the purified enzymes were generated. Appropriate PCR primers were derived from these peptides and employed for the screening of a *M. scorodonius* cDNA library. By means of primer walking, the entire cDNAs encoding MsP1 and MsP2 were cloned (accession numbers CS490657 and CS490663). The cDNA encoding MsP1 contained an open reading frame of 513 amino acids, while the mature protein (as deduced from the N-terminal Edman sequence) consisted of 458 amino acids (calculated MW ~49 kDA). Mature MsP2 was composed of 453 amino acids, corresponding to a calculated MW of ~48 kDa. The MsP1 and MsP2 encoding genes comprised 2093 bp and 2135 bp, respectively, and ten introns of 50 – 83 bp each.

**Table I. Volatile Degradation Products of Carotenes and Xanthophylls**

| *Substrate* | *Volatile degradation products* | *Yield [mol%]* |
|---|---|---|
| $\beta$-carotene | -ionone | 7.9 |
| | 5,6-epoxy-$\beta$-ionone | 1.3 |
| | dihydroactinidiolide | 7.0 |
| | $\beta$-cyclocitral | 1.5 |
| | 2-hydroxy-2,6,6-trimethylcyclo-hexanone | 2.5 |
| lutein | 3-hydroxy-$\alpha$-ionone | 11.0 |
| | 3-hydroxy-$\beta$-ionone | 6.3 |
| zeaxanthin | 3-hydroxy-$\beta$-ionone | 5.7 |
| | 3-hydroxy-$\beta$-cyclocitral | traces |
| violaxanthin, neoxanthin | 3-hydroxy-5,6-epoxy-$\beta$-ionone | 6.9<br>6.8 |
| lycopene | geranial | 1.3 |
| | 6-methyl-5-heptene-2-one | 7.2 |

In the amino acid sequence of MsP1, eight potential N-glycosylation sites and one potential O-glycosylation site were identified, while MsP2 contained four potential N-glycosylation sites and three potential O-glycosylation sites (*20*). By comparison to the data derived from MALDI-TOF experiments, a glycosylation degree of 23% was calculated for MsP1, and of 8% for MsP2. Glycosylation is typically not required for the catalytic activity of secreted fungal enzymes, but it increases the enzymes' thermostability and may also protect from degradation by peptidases (*22*).

Data bank homology searches with the amplified DNA and cDNA sequences of MsP1 and MsP2 indicated overall sequence homologies on the amino acid level of up to ~50% to so-called dye decolorizing peroxidases (DyP-type peroxidases). DyP-type peroxidases do not share sequence homologies with other fungal peroxidases like manganese peroxidases, lignin peroxidases, or versatile peroxidases, and they show unique reaction characteristics. They have thus not been grouped in the plant or animal peroxidase superfamilies (*23*). The DyP-type peroxidase superfamily currently comprises ~300 representatives from bacteria and fungi (http://peroxibase.toulouse.inra.fr/).

When various carotenes and xanthophylls were subjected to enzymatic degradation by MsP1, all of the substrates were readily degraded. Carbon-13 and carbon-10 norisoprenoids proved to be the main volatile reaction products in each case (Table I, Figure 5 (*24*);). To investigate if the entire tetraterpenoid compound is required for the enzymatic cleavage reaction, the apocarotenals *β*-apo-8′-carotenal and *β*-apo-12′-carotenal were additionally tested as substrates for the peroxidase catalysed carotenoid degradation (Figure 6).

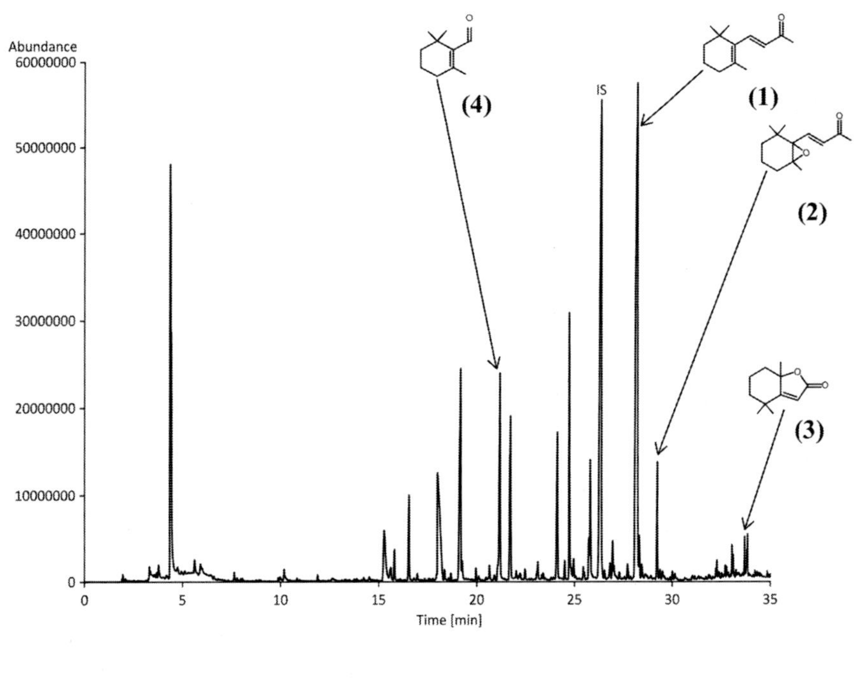


*Figure 5. Cleavage of β-carotene to flavor compounds by MsP1 catalysis (GC-MS-MS-chromatogram; HP-Innowax); (**1**) β-ionone, (**2**) 5,6-epoxy-β-ionone, (**3**) dihydroactinidiolide, (**4**) β-cyclocitral.*

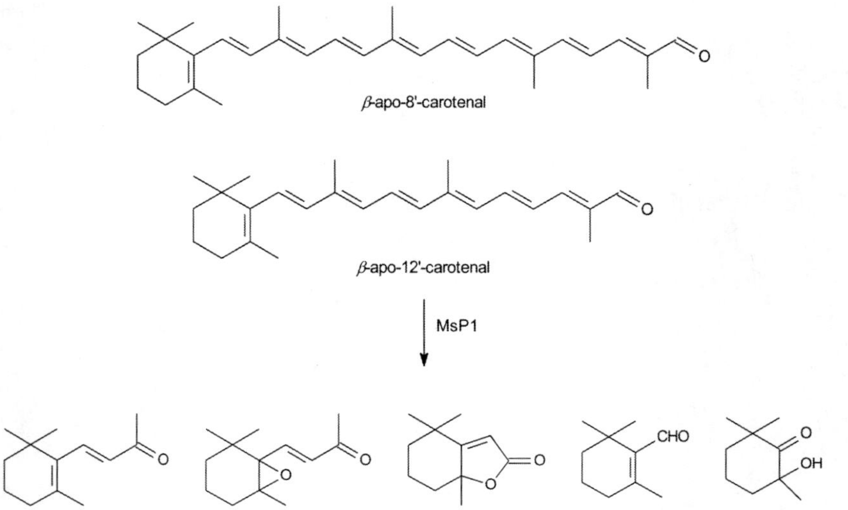


**(1)** [5.4 / 3.2] **(2)** [1.9 / 1.0] **(3)** [2.1 / 2.2] **(4)** [1.6 / 2.4] **(5)** [1.4 / traces]

*Figure 6. Release of norisoprenoids from apocarotenals by MsP1 catalysis; (**1**) – (**4**) cf. Figure 5; (**5**) 2-hydroxy-2,6,6-trimethyl-cyclohexanone; product yields are given in [mol%] for β-apo-8'-carotenal and β-apo-12'-carotenal, respectively. (Data modified after (24).)*

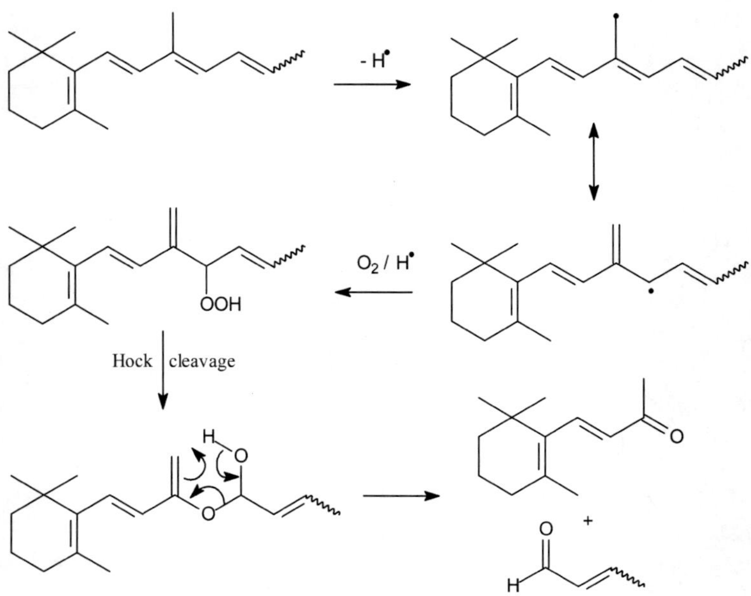


*Figure 7. Potential reaction mechanism of the versatile peroxidase catalysed oxidative cleavage of β-carotene to β-ionone and β-apo-10'-carotenal. (Modified after (25).)*

The spectrum of volatile norisoprenoids was comparable to that observed with $\beta$-carotene as substrate. In addition to the released volatile norisoprenoid compounds, the corresponding apocarotenals were tentatively identified by means of HPLC-MS and HPLC-DAD (Table II).

**Table II. Nonvolatile Degradation Products of Carotenes and Xanthophylls Tentatively Identified by HPLC-DAD and HPLC-MS Analyses[a]**

| *Substrate* | *[M + H⁺]* | *Tentative Identifiaction* |
|---|---|---|
| $\beta$-carotene | 311 | $\beta$-apo-14′-carotenal |
| | 351 | $\beta$-apo-12′-carotenal |
| | 377 | $\beta$-apo-10′-carotenal |
| | 553 | $\beta$-carotene-monoepoxide |
| | 553 | 5,6-epoxy-$\beta$-carotene |
| lutein / zeaxanthin | 327 | 3-hydroxy-apo-14′-carotenal |
| | 367 | 3-hydroxy-apo-12′-carotenal |
| | 393 | 3-hydroxy-apo-10′-carotenal |
| | 433 | 3-hydroxy-apo-8′-carotenal |
| | 585 | lutein / zeaxanthin epoxide |
| violaxanthin / neoxanthin | 343 | 3-hydroxy-5,6-epoxy-$\beta$-apo-14′-carotenal |
| | 383 | 3-hydroxy-5,6-epoxy-$\beta$-apo-12′-carotenal |
| | 409 | 3-hydroxy-5,6-epoxy-$\beta$-apo-10′-carotenal |

[a] (Data derived from (*24*)).

A second type of carotenoid degrading fungal heme peroxidases was detected in the secretome of *Pleurotus eryngii*, a close relative of the oyster mushroom *P. ostreatus*. In the original publication (*25*), *P. eryngii* was erroneously addressed as *Lepista irina*. From the supernatants of submerged grown *P. eryngii*, the $\beta$-carotene degrading enzyme was purified to electrophoretic homogeneity by means of fast protein liquid chromatography. *De novo* sequencing of tryptic peptides by HPLC-ESI-tandem mass spectrometry allowed for cloning of the encoding cDNA, and homology searches identified the target enzyme as a so-called versatile peroxidase (accession number AJ515245).

Fungal versatile peroxidases have become known as key enzymes of natural lignin degradation. They share catalytic properties of lignin peroxidases and manganese peroxidases. A broad spectrum of phenolic and non-phenolic aromatic compounds is oxidized either directly or indirectly through $Mn^{3+}$ formed from $Mn^{2+}$ as with manganese peroxidases (*26*). A potential mechanism of the versatile peroxidase catalysed carotenoid cleavage between C9 and C10 of the polyene chain involves the abstraction of a hydrogen atom from the allylic methyl group which yields a resonance-stabilized carbon radical. Hydroperoxides are formed intermediately through reaction with oxygen, and subsequent Hock cleavage yields two carbonyl compounds (Figure 7).

## Conclusions

DyP-type peroxidases and versatile peroxidases secreted by wood inhabiting basidiomycetes have been shown to efficiently degrade various carotenes and xanthophylls. Carbon-13 and carbon-10 norisoprenoids have been identified as the main volatile degradation products in each case. The basidiomycetous strains are comparatively easy to handle in submerged culture, and they are food-grade and non-toxic. The DyP-type peroxidase MsP1 proved to be highly stable up to temperatures of ~60 °C and a hydrostatic pressure of ~3 kbar. It could thus represent an interesting candidate for industrial norisoprenoid flavor production. Though the carotenoid substrates were degraded by the fungal enzymes nearly quantitatively under the experimental conditions, the overall yields of norisoprenoid aroma compounds were below 20 mol%. To further optimize the product yields, and to develop an adequate downstream-processing strategy remain ambitious aims for future research.

## Acknowledgments

Support of the work by the "Deutsche Forschungsgemeinschaft" (ZO 122/1-2 & INST 162/381-1 FUGG) is gratefully acknowledged.

## References

1. Werkhoff, P.; Bretschneider, W.; Güntert, M.; Hopp, R.; Surburg, H. *Z. Lebensm.-Unters. Forsch.* **1991**, *192*, 111–115.
2. Buttery, R. G.; Teranishi, R.; Ling, L. C. *Chem. Ind.* **1988**, *7*, 238.
3. Kumazawa, K.; Masuda, H. *J. Agric. Food Chem.* **1999**, *47*, 5169–5172.
4. Mayer, F.; Czerny, M.; Grosch, W. *Eur. Food Res. Technol.* **2000**, *211*, 272–276.
5. Wache, Y.; Bosser-DeRatuld, A.; Lhuguenot, J.-C.; Belin, J.-M. *J. Agric. Food Chem.* **2003**, *51*, 1984–1987.
6. Aziz, S.; Wu, Z.; Robinson, D. S. *Food Chem.* **1999**, *64*, 227–230.
7. Wache, Y.; Bosser-DeRatuld, A.; Belin, J.-M. In *Carotenoid-derived Aroma Compounds*; Winterhalter, P., Rouseff, R. L., Eds.; ACS Symposium Series 802; American Chemical Society: Washington, DC, 2002; pp 102−113.
8. MacTavish, H.; Davies, N. W.; Menary, R. C. In *Carotenoid-derived Aroma Compounds*; Winterhalter, P., Rouseff; R. L., Eds.; ACS Symposium Series 802; American Chemical Society: Washington, DC, 2002; pp 183−193.
9. Fleischmann, P.; Studer, K.; Winterhalter, P. *J. Agric. Food Chem.* **2002**, *50*, 1677–1680.

10. Suzuki, M.; Matsumoto, S.; Fleischmann, P.; Shimada, H.; Yamano, Y.; Ito, M.; Watanabe, N. In *Carotenoid-derived Aroma Compounds*; Winterhalter, P., Rouseff; R. L., Eds.; ACS Symposium Series 802; American Chemical Society: Washington, DC, 2002; pp 89−101.
11. Winterhalter, P.; Schreier, P. *Food Rev. Int.* **1995**, *11*, 237–254.
12. Schwartz, S. H.; Qin, X.; Loewen, M. C. *J. Biol. Chem.* **2004**, *279*, 46940–46945.
13. Booker, J.; Auldridge, M.; Wills, S.; McCarty, D. R.; Klee, H.; Leyser, O. *Curr. Biol.* **2004**, *14*, 1232–1238.
14. Schwartz, S. H.; Qin, X.; Zeevaart, J. A. D. *J. Biol. Chem.* **2001**, *276*, 25208–25211.
15. Huang, F-C.; Horváth, G.; Molnár, P.; Turcsi, E.; Deli, J.; Schrader, J.; Sandmann, G.; Schmidt, H.; Schwab, W. *Phytochemistry* **2009**, *70*, 457–464.
16. Fleischmann P.; Zorn H. In *Carotenoids - Vol. 4: Natural Functions*; Britton, G., Liaaen-Jensen, S., Pfander, H., Eds.; Birkhäuser: Basel, 2008; pp 341−366.
17. Nacke, C.; Schrader, J. *J. Mol. Catal. B: Enzym.* **2011**, *71*, 133–138.
18. Zorn, H.; Langhoff, S.; Scheibner, M.; Berger, R. G. *Appl. Microbiol. Biotechnol.* **2003**, *62*, 331–336.
19. Ben Aziz, A.; Grossman, S.; Ascarelli, I.; Budowski, P. *Phytochemistry* **1971**, *10*, 1445–1452.
20. Scheibner, M.; Hülsdau, B.; Zelena, K.; Nimtz, M.; de Boer, L.; Berger, R. G.; Zorn, H. *Appl. Microbiol. Biotechnol.* **2008**, *77*, 1241–1250.
21. Pühse, M.; Szweda, R. T.; Ma, Y.; Jeworrek, C.; Winter, R.; Zorn, H. *Biochim. Biophys. Acta* **2009**, *1794*, 1091–1098.
22. Sato, T.; Hara, S.; Matsui, T.; Sazaki, G.; Sajio, S.; Ganbe, T.; Tanaka, N.; Sugano, Y.; Shoda, M. *Acta Cryst.* **2004**, *60*, 149–152.
23. Sugano, Y. *Cell. Mol. Life Sci.* **2009**, *66*, 1387–403.
24. Zelena, K.; Hardebusch, B.; Hülsdau, B.; Berger, R. G.; Zorn, H. *J. Agric. Food Chem.* **2009**, *57*, 9951–9955.
25. Zorn, H.; Langhoff, S.; Scheibner, M.; Nimtz, M.; Berger, R. G. *Biol. Chem.* **2003**, *384*, 1049–1056.
26. Ruiz-Duenas, F. J.; Martinez, M. J.; Martinez, A. T. *Mol. Microbiol.* **1999**, *31*, 223–235.

## Chapter 14

# Design of Aqueous Micellar Reaction Systems for Aroma Production with Carotenoid Cleavage Dioxygenase

**Christoph Nacke,[1] Martin Schilling,[2] M. M. W. Etschmann,[1] and Jens Schrader*,[1]**

**[1]DECHEMA Research Institute, Theodor-Heuss-Allee 25, 60486 Frankfurt am Main, Germany**
**[2]Current address: Evonik Industries AG, Essen, Germany**
***E-mail: schrader@dechema.de.**

The use of carotenoid cleavage dioxygenases (CCD) as biocatalysts for aroma generation requires appropriate substrate delivery techniques. Non-ionic surfactants forming micelles represent a versatile tool to introduce hydrophobic carotenoids into aqueous reaction media. In this work, data obtained with CCD1 from *Arabidopsis thaliana* (AtCCD1) as model enzyme are presented. Previous results from studies with purified carotenoids as model substrates delivered in differently designed micelles are discussed together with new findings on the use of AtCCD1 for carotenoid cleavage out of micellized biomass as a technical substrate. The data presented illustrate the diversified impact of reaction parameters, such as surfactant type, carotenoid structure and source (i.e. purified or as component of biomass) as well as organic (co)solvents, on enzyme activity. New data reveal the possibility to directly cleave carotenoids out of carotenoid containing biomass. For this purpose crude AtCCD1 preparation was added to a homogenate of the carotenoid-rich cyanobacterium *Spirulina platensis* resulting in specific cleavage and high conversion yield.

# Introduction

Due to their key role in biosynthesis of bioactive apocarotenoids, carotenoid cleavage dioxygenases (CCD) have attracted tremendous interest from both fundamental and application oriented research after their genetic discovery at the end of the 1990s (*1–3*). Although much work has since been done to elucidate CCD substrate and product spectra (*4*, *5*) their molecular reaction mechanisms (*6*, *7*), their role in human carotenoid metabolism (*8*) or in phytohormone biosynthesis (*9*) and volatile products formation in plants (*10*), investigations towards a potential industrial application of CCD as biocatalysts for the production of flavor compounds have been rather scarce. From a biotechnological viewpoint, the two main hurdles for a technical application of CCD are the hydrophobicity of the substrates, making them virtually insoluble in conventional aqueous systems, and the difficulties associated with CCD expression in microbial hosts leading to low enzyme yields. Whereas the latter aspect is still a severe bottleneck (*11*, *12*), significant progress has been achieved to improve the in vitro performance of CCD by tailoring substrate delivery. Here we present data obtained with CCD1 from *Arabidopsis thaliana* (AtCCD1) and differently designed carotenoid-loaded micelles in aqueous model reactions to produce the violet-like flavor and fragrance compound β-ionone (*11*, *13*). These approaches deepened our understanding of substrate recruitment mechanisms in vesicle-based reaction systems and will be beneficial to other enzymatic reactions with hydrophobic substrates as well.

The production of 'natural' flavouring substances according to the European regulation on flavors (1334/2008/EC) requires the use of a substrate of vegetable, animal or microbiological origin for microbiologcal processes (*14*). Since the cost of separation and purification of natural carotenoids is high, the use of carotenoid containing biomass as source for the required natural carotenoids would be economically very attractive. Therefore, first results illustrating the transferability of CCD based aroma formation from model to technical substrates are presented, exemplified by the use of cyanobacterial *Spirulina platensis* biomass.

# Materials and Methods

## Reagents and Materials

Carotenoids were purchased from Fluka Chemicals, Buchs, Switzerland. Surfactants were purchased from Sigma–Aldrich GmbH, Steinheim, Germany. Other chemicals were purchased at analytical grade purity from Roth Chemicals, Karlsruhe, Germany. Water was obtained from a NANOpure UV water purification system (Thermo Fisher Scientific, NC, U.S.A.).

## Transformation, Expression, and Preparation of Cell Extracts

AtCCD1 was amplified as described by Schwartz et al. (*1*). Expression of the native, non-tagged form of AtCCD1 was achieved by cloning into pET29a as described by Schilling et al. (*11*), however with a different set of primers: Forward: 5′-T ACA TTA ATG GCG GAG AAA CTC AGT G-3′; Reverse: 5′-TA CAT TAA TGC TTA TAT AAG AGT TTG TTC C-3′. In order to express a GST-AtCCD1 (an N-terminal fusion of AtCCD1 with glutathione S transferase to increase the solubility of the fusion protein), the gene was also cloned behind the tac promoter into pGEX-4T-1 (Amersham Biosciences) via the BamHI/EcoRI restriction sites. Sequencing confirmed orientation and correct ligation.

*E. coli* BL21 DE3 cells (Novagen, U.K.) were heat-shock transformed with pGEX-4T-1-AtCCD1 or pET29a-AtCCD1 and expression was verified by SDS-PAGE. Protein expression was conducted at 20° C and otherwise as described by Schilling et al. (*11*). After sonication, 1.85mM Triton X-100 was added for optimum protein solubilisation. After mixing and incubationon on ice for 2min, cell debris was removed by 25 min of centrifugation at 5600×g and 4° C. In all experiments except for the data displayed in Figure 2, this crude lysate was used as enzyme solution. The data shown in Figure 2 was obtained using affinity purified GST-AtCCD1. Purification was performed as described by Schilling et al. (*11*).

Protein concentrations were determined by bicinchoninic acid (BCA) assay. An eight-point calibration using bovine serum albumine as a reference substance was performed with every measurement. The molar concentrations of enzymes were calculated by multiplication of the measured protein concentration of the crude extract or purified fraction with the purity factor. The purity factor was determined by relating the optical density of the respective enzyme band to the sum of the optical density of all protein bands. To obtain molar concentrations, the mass concentrations were divided by the respective molecular weights of the (fusion) proteins (62 and 87 kD for native AtCCD1 and GST-AtCCD1, respectively).

## Carotenoid Cleavage Activity Assays

Carotenoid cleavage reactions were measured photometrically by detecting the decrease in carotenoid absorbance during carotenoid cleavage reactions performed in lid-covered polystyrene microtiter plates as described in the literature (*11*, *13*). Except for the data presented in Figure 1 and 2, we used the highest molar substrate to surfactant ratio permitting reliable solubilization of all tested carotenoids in all tested surfactans without risk of precipitation. Determinations of reaction velocities were usually done in triplicate. For data shown in Figure 2C, six parallel measurements of the reaction at 0 and 15% (v/v) ethanol were performed to calculate the degree of activation.

### Conversion of Carotenoid Rich Biomass

*S. platensis* biomass was disrupted in 50 mM Tris pH 8.5 by sonication in a 20°C water bath with a total energy input of 2.4 kJ/mL at intervals of 0.5 s sonication and 0.5 s breaks in-between. After cell disruption, (co)solvents were added at 27 % (v/v), resulting in a (co)solvent concentration in the assay of 10 % (v/v). Micellization was conducted for 60 min in the presence of the (co)solvents by addition of 8 mM Tween 40 at 40°C and 1400 rpm in a shaker incubator. Reactions were conducted in closed 2 mL polypropylene vessels for 14 h at 30°C and 1000 rpm in shaker incubators. Biomass concentration was 125 g/L with 10 g/L Tween 40 and 50 mM Tris pH 8.5 as reaction buffer. Negative controls were run as described for experiments with purified carotenoid substrates. Cosolvents for improved micellization were added at 27 % (v/v) during micellization, leading to a concentration of 10 % (v/v) during the enzymatic reaction.

## Results and Discussion

### Impact of Surfactant Type on Carotenoid Cleavage out of Micelles

Different structurally related non-ionic surfactants (Tween 20, 40, 60, 80 and Triton X-100) were investigated for their ability to deliver carotenoid substrates of different hydrophobicity to native AtCCD1. Experiments were carried out with native AtCCD1 in microtiter plates with a substrate concentration of 4.8 µM and a surfactant concentration of 16 mM which corresponded to a value of more than 10% above the critical micelle concentration in all cases. These substrates were chosen because they cover a broad range of different hydrophobicities as represented by their octanol/water partition coefficient log P (o/w) (Figure 1).

The maximum reaction rates obtained with the different surfactant - substrate configurations are shown in Figure 1. Reaction rates using different surfactants can only be compared within one experimental series (the same substrate). A comparison between different substrates is not feasible as the enzyme solution used in the different experiments was stored for different time spans.

Figure 1 illustrates that depending on the substrate tested a different spectrum of surfactants turned out to form micelles permitting optimal substrate delivery and conversion by the enzyme. Importantly, for the cleavage of β-carotene the surfactant Tween 40 worked best. The results reveal that only highly oxyfunctionalized carotenoids such as astaxanthin and apocarotenoids can be delivered efficiently to AtCCD1 using Triton X-100 micelles. In contrast, the less oxyfunctionalized zeaxanthin and the fully non-polar β-carotene are cleaved only very poorly out of Triton X-100 micelles. This is an important result as hitherto Triton X-100 had not only been used routinely for the biochemical characterization of CCD, usually in combination with the readily cleavable model substrate 8′-apo-β-caroten-8′-al, but also to investigate the performance towards the economically more interesting substrate β-carotene. Consequently, for in vitro β-carotene cleavage, no significant yields with CCD had been described beforehand. Thus, data obtained with micelle-based reaction systems (*2*, *4*, *15–17*) should be critically reviewed regarding the interpretation of substrate preferences of the respective CCD used. The specific surfactant type used significantly influences the substrate preference of CCDs.

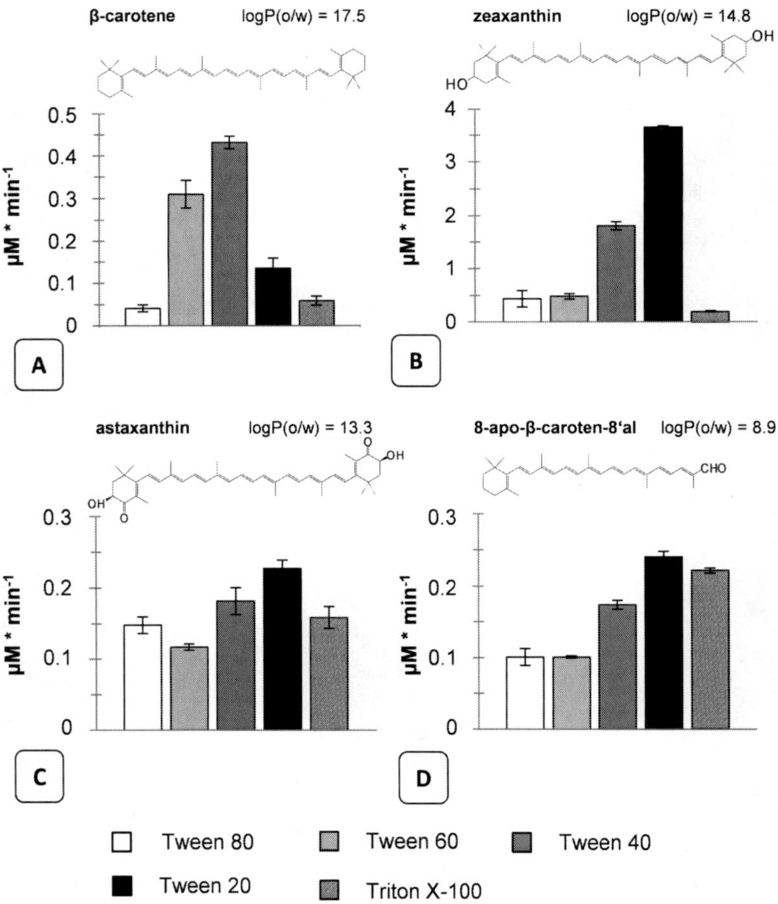


*Figure 1. Comparison of reaction velocities obtained using different nonionic surfactants for micelle-based delivery of four typical carotenoid substrates to native AtCCD1. β-carotene (A), zeaxanthin (B), astaxanthin (C) and 8'-apo-β-caroten-8'-al (D). Substrate concentration: 4.8 µM. Surfactant concentration: 16 mM. Adapted from Reference (13), Copyright 2012, Elsevier.*

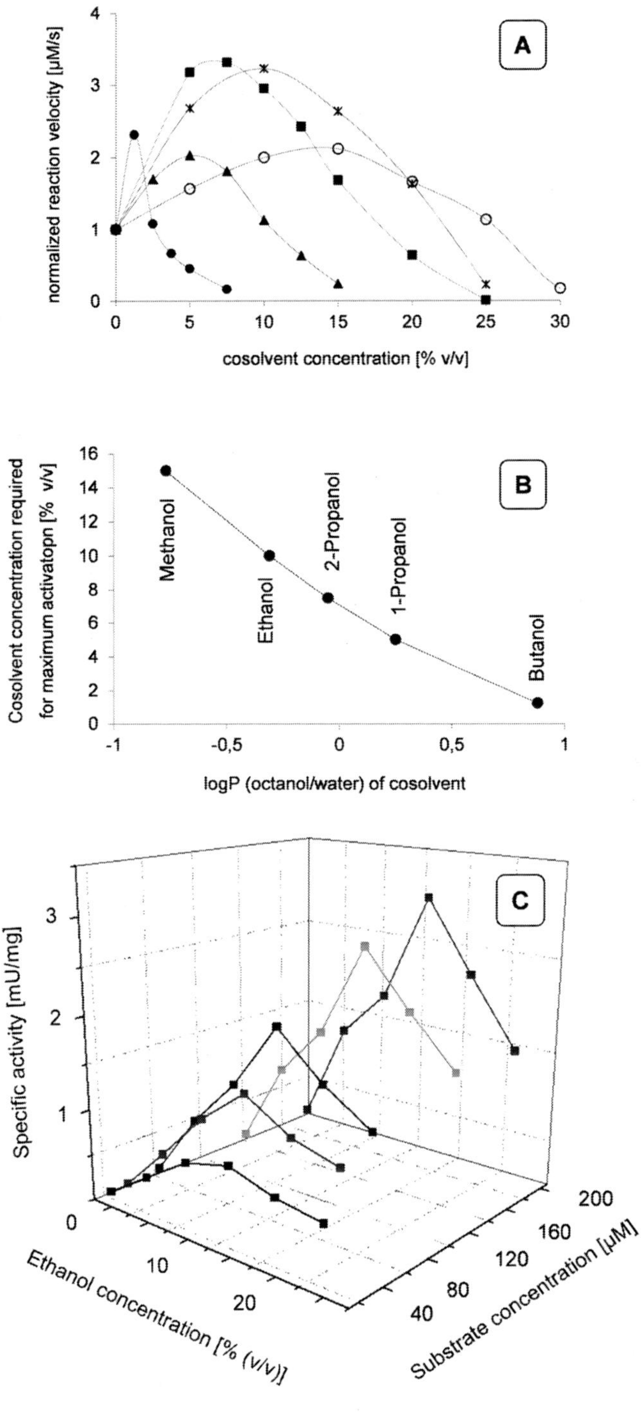
A
normalized reaction velocity [µM/s]
cosolvent concentration [% v/v]
B
Cosolvent concentration required for maximum activatopn [% v/v]
logP (octanol/water) of cosolvent
Methanol
Ethanol
2-Propanol
1-Propanol
Butanol
C
Specific activity [mU/mg]
Ethanol concentration [% (v/v)]
Substrate concentration [µM]

*Figure 2. (A) Reaction velocity at different concentrations of water-soluble cosolvents with hydrophobicities between logP = -0.7 and logP = +0.9 . (•) 1-butanol, (▴) 1-propanol, (■) 2-propanol, (x) ethanol, (○) methanol. Reaction system: 1% (w/v) Triton X-100, 63 μM 8'-apo-β-caroten-8'-al. (B) Cosolvent concentration required for maximum activation versus octanol-water partitioning coefficient logP of the tested cosolvents. (C) Dependence of GST-AtCCD1 activation by cosolvents on substrate concentration. The molar substrate to surfactant ratio was 0.008 in all measurements. Substrate concentrations: 189 μM, 126 μM, 50 μM, 25 μM, 12 μM.*

## Impact of Water-Miscible Organic Solvents on Carotenoid Cleavage out of Micelles

To investigate the effect of water-miscible organic solvents on enzyme activity in micellar reaction systems, the influence of a series of aliphatic alcohols with increasing chain lengths and log P values was determined in microplate assays with purified AtCCD1. Figure 2A illustrates the dependence of the enzyme activity on type and concentration of the organic solvent. Up to a specific concentration, an increase in activity was observed for all alcohols investigated, although the degree of activation as well as the concentration with the highest activity increase varied between the solvents. While the extent of activation was comparable for all alcohols, the optimum cosolvent concentration shifted towards lower values with increasing log P values (Figure 2B), and inactivation was observed when the cosolvent concentration was further raised (Figure 2B). Consequently, activation and inactivation in the micellar system seems to be mainly dependent on the hydrophobicity of the resulting water/alcohol mixture. While protein denaturation can be taken as the main reason for inactivation effects at increased organic cosolvent concentrations (*18*), the activation effect observed at lower alcohol concentrations may be rationalized by both an effect on enzyme structure / active site hydration and an effect on micellar structure known from other enzymes (*19–23*). To underpin our hypothesis that a micelle restructuring may play a dominant role in activation of carotenoid cleavage by CCD, the effect of various ethanol concentrations was monitored over a range of different substrate concentrations at a constant substrate/surfactant ratio with GST-AtCCD1. Optimal ethanol concentration for this fusion enzyme was determined to be 15% (v/v), and an increasing activation up to an 18-fold (±0.7) activation of the enzyme activity was found for the experiment with the highest substrate and surfactant concentration tested (Figure 2C). The fact that the degree of activation increased with increasing substrate and detergent concentrations was in accordance with our hypothesis, that substrate accessibility or micellar structure play a major role in enzyme activation while a significant impact of the low concentrated alcohol on the enzyme can be exluded, as this would have led to a constant activation degree independent on the substrate concentration. In addition, alcohol stability of the enzyme seems to be influenced by the GST-tag fused to AtCCD1 as illustrated by the increased optimal ethanol concentration of the fusion enzyme (Figure 2C) compared to the native AtCCD1 (Figure 2A).

## Impact of Surfactant Structure on Carotenoid Cleavage out of Micelles

To gain a better understanding of the molecular mechanisms involved in the substrate delivery from the micelle to the enzyme, the role of the surfactant molecular structure was investigated systematically with the Tween surfactant series. The different Tweens only differ in the length and saturation of their hydrophobic chains (Figure 3A). Native AtCCD1 was used in order to exclude possible effects by a fusion tag together with 8′-apo-β-caroten-8′-al and zeaxanthin as substrates and ethanol as cosolvent. Figure 3B and 3C illustrate that for both substrates the cosolvent concentration for maximum activation increased with the length of the hydrophobic chain of the surfactant. The non-saturated carbon-carbon bond in Tween 80 dramatically reduced the ability of the formed micelles to deliver the substrates to the enzyme.

These data supported our hypothesis that the accelerated conversion of hydrophobic substrates from non-ionic surfactant micelles is a result of a restructuring of the micelles caused by the organic cosolvent. The presence of water-soluble organic cosolvents causes an increase of the critical micelle concentration (CMC) by reducing the hydrophobic interactions between surfactant molecules (*24*). Due to the lowered CMC, the aggregation number Nagg of the micelles (the number of surfactant molecules per micelle) is reduced. Consequently, the distance between the substrate in the hydrophobic micelle core and the micelle surface is shortened (*25*). In accordance with our results on liposome based carotenoid delivery (*26*), we postulate that an ideal distance of the carotenoid from the micelle surface exists for optimum conversion by AtCCD1. Surfactants with longer hydrophobic chains form micelles of larger diameters. Their diameter therefore has to be reduced more by the addition of more cosolvents until the ideal distance between substrate and micelle surface is obtained. Interestingly, an unsaturated double bond in the aliphatic side chain of the surfactant caused a total loss of carotenoid delivery capacity of the respective micelle. Available data about critical micelle concentration, aggregation number and micelle diameter (*27*) do not indicate any other micelle property as an explanation for the observed effect. Possibly, the hydrophobic core of micelles formed by this surfactant is less homogenous and less pronounced due to steric hindrance of the bent hydrophobic chains, since commercial Tween 80 is a racemic mixture of cis and trans isomers.

## Cleavage of Carotenoids out of Carotenoid-Rich Biomass as a Technical Substrate

Using technical substrates (carotenoid containing biomass) instead of purified carotenoids as substrate for the proposed process would provide both economical and ecological advantages due to the low cost of biomass and the low upstream processing required in comparison to purified carotenoids. In contrast to other enzymes proposed for carotenoid cleavage into flavor and fragrance compounds such as the peroxidase of *Lepista irina* (*32*), AtCCD1 is selective to carotenoids and cleaves them in a highly regiospecific manner. Working with very complex substance mixtures such as disrupted cells, the specificity of AtCCD1 would avoid the formation of undesired side products.

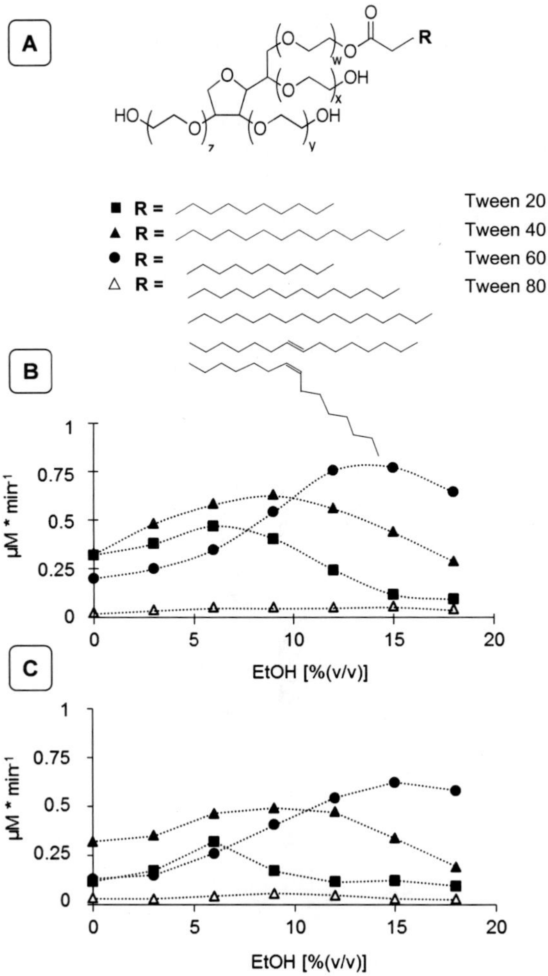


*Figure 3. Influence of surfactant molecular structure (A) on the ethanol concentration required for maximum acceleration of carotenoid cleavage by native AtCCD1. Substrate: 48 μM 8'-apo-β-caroten-8'-al (B) or 48 μM Zeaxanthin (C). Surfactant concentration: 160mM. Adapted, in parts, from Reference (13), Copyright 2012, Elsevier.*

In order to point out its technical potential, we decided to investigate the general usability of cyanobacterial biomass as technical substrate. Though obtained from preliminary experiments carried out on small scale, the data already provide valuable information on the use of technical carotenoid sources in the future.

The passage of carotenoids from the membrane fragments and emulsion droplets formed by cell disruption into mixed micelles has already been investigated in research concerning the intestinal absorption of carotenoids. It was shown that especially the strongly hydrophobic carotenoids such as β-carotene are difficult to mobilize from their natural storage locations into mixed micelles (*28*). Therefore, the influence of organic solvents on the micellization and bioconversion of carotenoids from *Spirulina platensis* biomass was investigated using the water-soluble organic solvent ethanol and the water insoluble biocompatible solvents oleyl alcohol and butyl stearate. The addition of solvent was performed after cell disruption and before the micellization step. Reactions were only conducted for fourteen hours, therefore conversion did not represent the maximum values achievable. Negative controls were used to compensate for carotenoid loss by autoxidation or non-specific side reactions with components of the reaction solution. For β-carotene, an improved conversion could be achieved with all solvents tested (Figure 4A). For the less hydrophobic xanthophylls zeaxanthin, only the water-soluble cosolvent ethanol led to an improved conversion (Figure 4B).

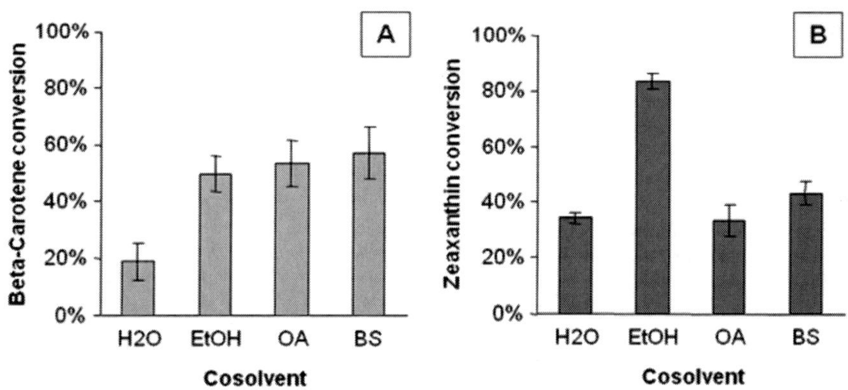


*Figure 4. Conversion of the carotenoids β-carotene (A) and zeaxanthin (B) from Spirulina platensis biomass in presence of different water miscible and non water miscible solvents during micellization. Solvent concentration: 27 %(v/v) during micellization, 10 %(v/v) during reaction. H2O: Water, EtOH: Ethanol, OA: Oleyl alcohol (cis-9-Octadecen-1-ol), BS: Butyl stearate. The 100% value represents the total carotenoid content of the algae and was determined via organic extraction.*

The different results obtained for the fully non-polar β-carotene and the partially polar zeaxanthin indicate that depending on substrate structure, two different effects can occur during micellization in presence of organic (co)solvents. In case of the fully non-polar β-carotene, all types of solvents tested led to increased conversion. This increased conversion is most likely the result of a larger share of substrate that was mobilized from its membraneous storage compartments into vesicles in which it is accessible to the dioxygenase enzyme. In contrast, the partially polar zeaxanthin is only cleaved at increased yield if the water-miscible cosolvent ethanol is used while the non water miscible biocompatible solvents oleyl alcohol and butyl stearate did not cause any effect. The data indicate that an increased reaction velocity rather than improved micellization of the substrate is the effect observed for zeaxanthin because improved micellization would, according to theory, be observable with all three types of solvent tested. The hypothesis is supported by the work of Tyssandier and coworkers (*28*) showing that xanthophylls such as zeaxanthin are micellized from biomass much more readily than the fully non-polar carotenoids, even without the use of (co)solvents. The addition of ethanol during micellization and reaction most likely causes a restructuring of the micelles leading to an increased reaction velocity of the CCD as described in our earlier work for a structurally similar apocarotenoid substrate in a biomass free micelle based reaction system (*11*).The improved mobilization of β-carotene from cyanobacterial biomass by two biocompatible non water miscible solvents is of particular interest since this technique could permit a continuous extraction of β-carotene from a viable cyanobacteria culture, therefore combining a continuous carotenoid bioconversion with continuous substrate supply. Such a continuous biocompatible 'milking' of carotenoids was already described for green algae by Mojaat and coworkers (*29*), indicating the technological potential of this method.

## Conclusions

The micellar delivery of carotenoids is a versatile tool for substrate delivery to carotenoid cleavage enzymes in aqueous reaction media. Through choice of surfactant type and organic cosolvent addition, the micellar reaction system can be tailored for optimum cleavage of selected carotenoid substrates. Typical volatile by-products, such as dihydroactinidiolide, observed after unspecific, lipoxygenase-catalyzed co-oxidative cleavage (*30*, *31*) or cleavage with directly acting fungal peroxidases (*32*), were not observed with the highly regiospecific AtCCD1. Almost theoretical yields (> 90 mol% β-ionone) were obtained when spectrophotometrically quantified substrate cleavage and GC based analysis of product formation were compared (*11*). This emphasizes the potential of the CCD family for biocatalytic applications.

Based on the improved understanding of micellar substrate delivery it was possible to step forward to the direct conversion of carotenoids out of biomass as a technical substrate, the ultimate goal for a low-cost production of β-ionone and other carotenoid-derived volatiles as natural aroma compounds. Here the superiority of CCDs compared to non-specifically cleaving enzymes can be harnessed to its full extent as, despite the presence of a multitude of organic molecules, only the target carotenoids are cleaved and the desired C13 norisoprenoids are formed as the sole products. The investigation of biomass other than cyanobacteria in the future, such as extraordinarily carotenoid-rich algal of fungal biomass or even side-streams of the food industry, e.g. carrot pomace from juice, may grant access to natural C13 norisoprenoid aroma compounds in an economically attractive way.

# References


1. Schwartz, S. H.; Tan, B. C.; Gage, D. A.; Zeevaart, J. A. D.; McCarty, D. R. *Science (Washington, DC, U. S.)* **1997**, *276*, 1872–1874.
2. Schwartz, S. H.; Qin, X.; Zeevaart, J. A. D. *J. Biol. Chem.* **2001**, *276*, 25208–25211.
3. Kloer, D. P.; Schulz, G. E. *Cell. Mol. Life Sci.* **2006**, *63*, 2291–2303.
4. Huang, F.-C.; Horváth, G.; Molnár, P.; Turcsi, E.; Deli, J.; Schrader, J.; Sandmann, G.; Schmidt, H.; Schwab, W. *Phytochemistry* **2009**, *70*, 457–464.
5. Walter, M. H.; Strack, D. *Nat. Prod. Rep.* **2011**, *28*, 663–692.
6. Leuenberger, M. G.; Engeloch-Jarret, C.; Woggon, W.-D. *Angew. Chem.* **2001**, *113*, 2683–2687.
7. Schmidt, H.; Kurtzer, R.; Eisenreich, W.; Schwab, W. *J. Biol. Chem.* **2006**, *281*, 9845–9851.
8. von Lintig, J.; Vogt, K. *J. Nutr.* **2004**, *134*, 251S–256S.
9. Alder, A.; Jamil, M.; Marzorati, M.; Bruno, M.; Vermathen, M.; Bigler, P.; Ghisla, S.; Bouwmeester, H.; Beyer, P.; Al-Babili, S. *Science (Washington, DC, U. S.)* **2012**, *335*, 1348–1351.
10. Auldridge, M. E.; McCarty, D. R.; Klee, H. J. *Curr. Opin. Plant Biol.* **2006**, *9*, 315–321.
11. Schilling, M.; Patett, F.; Schwab, W.; Schrader, J. *Appl. Microbiol. Biotechnol.* **2007**, *75*, 829–836.
12. Nacke, C.; Schilling, M.; Schrader, J. *J. Biotechnol.* **2009**, *144*, 268–271.
13. Nacke, C.; Schrader, J. *J. Mol. Catal. B: Enzymol.* **2012**, *77*, 67–73.
14. Demyttenaere, J. C. R. *Flavour Frag. J.* **2012**, *27*, 3–12.
15. Huang, F.-C.; Molnár, P.; Schwab, W. *J. Exp. Bot.* **2009**, *60*, 3011–3022.
16. Schwartz, S. H.; Qin, X.; Loewen, M. C. *J. Biol. Chem.* **2004**, *279*, 46940–46945.
17. Rubio, A.; Rambla, J. L.; Santaella, M.; Gomez, M. D.; Orzaez, D.; Granell, A.; Gomez-Gomez, L. *J. Biol. Chem.* **2008**, *283*, 24816–24825.
18. Schiffer, C. A.; Doetsch, V. *Curr. Opin. Biotechnol.* **1996**, *7*, 428–432.
19. Wu, Y.; Roberts, M. F. *Biochemistry* **1997**, *36*, 8514–8521.
20. Wehbi, H.; Feng, J.; Roberts, M. F. *Biochim. Biophys. Acta, Biomembr.* **2003**, *1613*, 15–27.
21. Meziani, A.; Touraud, D.; Zradba, A.; Pulvin, S.; Pezron, I.; Clausse, M.; Kunz, W. *J. Phys. Chem. B* **1997**, *101*, 3620–3625.
22. Schirmer, C.; Liu, Y.; Touraud, D.; Meziani, A.; Pulvin, S.; Kunz, W. *J. Phys. Chem. B* **2002**, *106*, 7414–7421.
23. Tomsic, M.; Bester-Rogac, M.; Jamnik, A.; Kunz, W.; Touraud, D.; Bergmann, A.; Glatter, O. *J. Colloid.*
24. Carnero Ruiz, C.; Molina-Bolívar, J. A.; Aguiar, J.; MacIsaac, G.; Moroze, S.; Palepu, R. *Colloid Polym. Sci.* **2003**, *281*, 531–541.
25. Bhattacharya, S. C.; Palepu, R. M. *J. Surf. Sci. Technol.* **2004**, *20*, 159–177.
26. (a) Nacke, C.; Schrader, J. *J. Mol. Catal. B: Enzymol.* **2011**, *71*, 133–138. (b) Nacke, C.; Schrader, J. *Interface Sci.* **2006**, *294*, 194–211.
27. Glenn, K. M.; Moroze, S.; Palepu, R. M.; Bhattacharya, S. C. *J. Dispersion Sci. Technol.* **2005**, *26*, 79–86.

28. Tyssandier, V.; Lyan, B.; Borel, P. *Biochim. Biophys. Acta, Mol. Cell. Biol. Lipids* **2001**, *1533*, 285–292.
29. Mojaat, M.; Foucault, A.; Pruvost, J.; Legrand, J. *J. Biotechnol.* **2008**, *133*, 433–441.
30. Aziz, S.; Wu, Z.; Robinson, D. S. *Food Chem.* **1999**, *64*, 227–230.
31. Wache, Y.; Bosser-DeRatuld, A.; Ly, H. M.; Belin, J.-M. *J. Mol. Catal. B: Enzymol.* **2002**, *19-20*, 197–201.
32. Zorn, H.; Langhoff, S.; Scheibner, M.; Nimtz, M.; Berger, R. G. *Biol. Chem.* **2002**, *384*, 1049–1056.

# Chapter 15

# In Situ Product Recovery of β-Ionone by Organophilic Pervaporation

**Maria M. W. Etschmann,* Christoph Nacke, Robert Walisko, and Jens Schrader**

**DECHEMA Research Institute, Theodor-Heuss-Allee 25, 60486 Frankfurt am Main, Germany**
***E-mail: etschmann@dechema.de.**

Volatile carotenoid cleavage products are of great commercial interest as aroma compounds. While tremendous advancements in understanding genetics and biochemistry of enzymatic carotenoid cleavage have been achieved during the last decade, investigations towards a future biotechnological application of these enzymes have been rare to date. Here we present organophilic pervaporation (OP) as a promising membrane technique for *in-situ* removal and selective enrichment of the volatile and hydrophobic β-ionone from aqueous in vitro reaction systems with *Arabidopsis thaliana* carotenoid cleavage dioxygenase 1 (AtCCD1). This membrane technique uses homogeneous silicone membranes for the selective enrichment of hydrophobic compounds from aqeuous solutions, here in enzymatic reaction systems. The resulting permeate is a highly concentrated aqueous solution of the product which facilitates downstream processing. Successful proofs-of-concept for coupling 8′-apo-β-caroten-8′-al and beta-carotene cleavage with OP will be given which also reveal the need for further optimization.

# Background

## The Aroma Compound β-Ionone

β -Ionone smells characteristically of violets, whose essential oil is to date the source for the natural compound, making it a very precious and expensive product. The worldwide annual production of cheap synthetic β-ionone is between 4,000 and 8,000 tons. It can be concluded that a market for the natural product definitely does exist. A biotechnological route for natural β-ionone would also make way for its use as a flavor compound. As consumers prefer "natural" flavors over synthetic ones, β-ionone is rarely used in flavor compositions because of the prohibitively high cost of the natural product (*1*). It has been known for some time that natural β-ionone is accessible by cleaving carotenoids (*2*) however, this route is not exploited commercially up to now.

## Principle of Organophilic Pervaporation

Pervaporation in general is a membrane process characterized by the properties of the membrane. Hydrophilic pervaporation is widely applied for the de-watering of organic solvent. The organophilic version however, constitutes a niche market (*3*) although it is very well suited for the in-situ recovery of volatile organic aroma compounds. Figure 1 shows the principle of organophilic pervaporation.

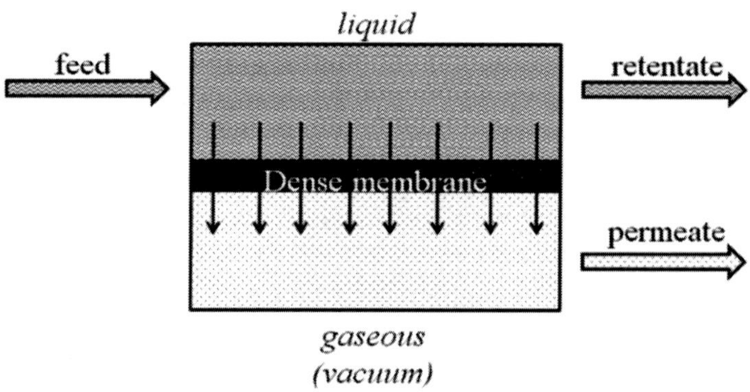


*Figure 1. Principle of organophilic pervaporation.*

The membrane acts as a physical barrier between feed and permeate side. The separating effect is due to a preferred interaction of the organic components in the feed with the hydrophobic membrane. While the liquid feed stream, i.e. culture medium flows over the membrane, organophilic components adsorb to the membrane and diffuse through it. As there are no pores in the membrane there is no risk of clogging the membrane even when culture broth including cells is circulated. As no cell retention is necessary pervaporation is ideal for whole cell processes (*4*, *5*). The coupling of enzymatic processes with pervaporation has not been published to this date and is an innovation of the presenting workgroup.

The pervaporation membrane retains most of the components of the liquid feed stream which is therefore called retentate. Only volatile and organophilic substances can permeate through the membrane leading to the term permeate for the condensed product vapors. While the pressure on the retentate side is ambient, underpressure is applied on the permeate side to support vaporization. The phase change from liquid to gaseous is one of the characteristics of pervaporation, the necessary vaporization enthalpy is drawn from the feed stream.The actual driving force of the process, however, is a gradient in chemical potential of the product between the retentate and the permeate side, manifested in a concentration gradient and gradient of vapour pressure, respectively (*6*).

## Performance Figures of Pervaporation

An important parameter to assess the efficiency of a pervaporation process is the flux. It describes the amount of permeate that passes through the membrane per area and time unit. For mixtures of components the overall flux is the sum of the partial fluxes of the single components.

For comparison and evaluation of separation processes the enrichment factor is used, which is the concentration ratio of the compound in feed and permeate.

## Composition of the Pervaporation Membrane

In all membrane processes the thickness of membrane affects the flux. Usually flux and membrane thickness are inversely proportional, therefore the membrane should be as thin as it possibly can be produced. For sheet membranes the compromise between membrane thinness and mechanical stability is a composite membrane. An inert fleece with wide pores is used as mechanically stable support and coated with a polymer film serving as the active layer, which can then be as thin as some micrometers.

## Characterization of the Pervaporation Membranes Used

The organophilic components in the retentate adsorb onto the hydrophobic membrane, diffuse through the polymer and are desorbed on the permeate side. Due to the underpressure they undergo a phase change into the gaseous state. The permeate gasstream is condensed in freeze traps cooled by liquid nitrogen. Upon thawing it is a concentrated aqueous-ethanolic solution of products. The adsorption velocity of the component that needs to be removed is dependent on the concentration of the respective component in the feed boundary layer. To keep the boundary layer as thin as possible and achieve concentration polarisation the flow over the pervaporation membrane needs to be as fast as possible but at least turbulent.

# Results

## Membrane Screening

Organophilic pervaporation is an niche market and in contrast to hydrophilic membranes the acquisition of the membranes for research purposes is a challenge in itself. If organophilic membranes are available, suppliers often only sell them in bulk quantities. It is difficult to predict the suitability of a membrane for the removal of the specific product. Therefore different membrane materials have to be screened for each product with the constraint of what is available in small amounts.

Table 1 shows the membranes and their most important parameters that were screened for *in-situ* product removal of β -ionone.

**Table 1. Organophilic Membranes Screened for the in Situ Removal of β-Ionone**

| *Name* | *Thickness of the active layer* | *Polymer of the active layer* | *Supplier* |
|---|---|---|---|
| 2420G/V-1 | 2μm | PDMS poly-dimethylsilox-ane | GMT, Rheinfelden, Germany |
| Pervap 4060 | 10 μm | polyacrylnitrile | Sulzer Chemtech, Allschwil, Switzerland |
| POMS/PEI | 10 μm | POMS poly-octylmethyl-siloxane | GKSS, Geesthacht, Germany |

For the screening an aqueous model solution of 100 mg/L β-ionone with 10% (v/v) ethanol as cosolvent was circulated in a standard pervaporation test module (GKSS, Geesthacht, Germany) which was equipped with 35 $cm^2$ of the respective membrane material. The freeze traps were changed every hour and substrate and product concentrations determined by HPLC.

When permeate concentrations are plotted over retentate concentrations and a linear regression is forced through zero, the resulting slope corresponds with the enrichment factor of the respective membrane for the specific product.

According to the enrichment factor the POMS membrane is suited best for the removal of β-ionone from the reaction solution. However, it also removes high amounts of ethanol. In general this is desirable as it leads to flux coupling and ameliorates permeation of the products. It needs to be kept in mind that in this special case ethanol plays an important role for the substrate delivery in the micellar reaction system and ethanol removal is therefore undesirable (Nacke et al., in this book). Pervap 4060 offers the advantage of leaving more ethanol behind in the retentate (data not shown) and is therefore the membrane of choice albeit with a smaller enrichment factor.

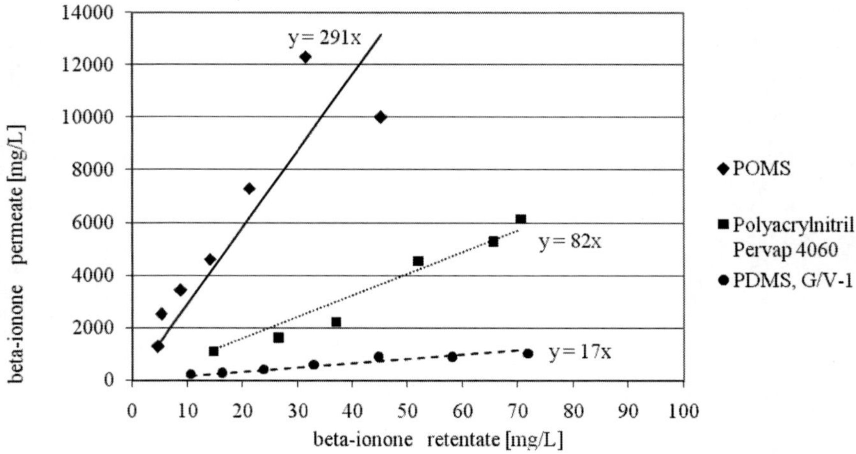


*Figure 2. Product concentrations in retentate and permeate during the pervaporation of an aqueous-ethanolic model solution of β-ionone with different membranes.*

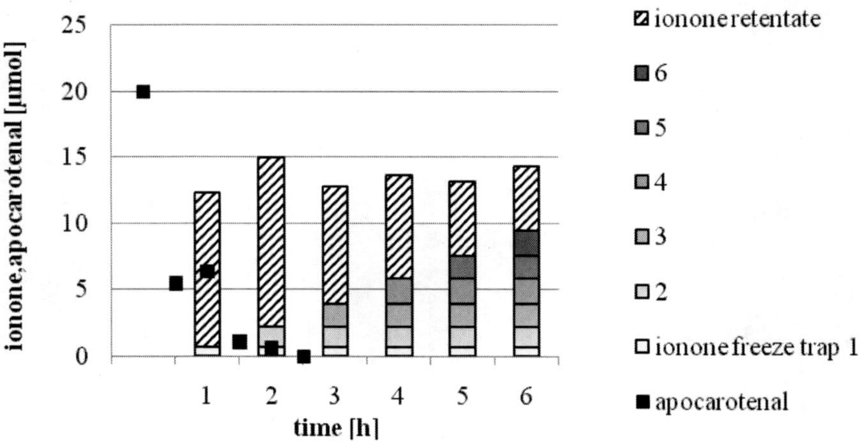


*Figure 3. Substrate amount in the retentate and product amounts in retentate and permeate during the enzymatic cleavage of 8′-apo-β-caroten-8′-al by AtCCD1 with product removal by organophilic pervaporation.*

**Enzymatic Cleavage of the Model Substrate 8′-Apo-β-caroten-8′-al**

Cell extract from *E.coli* pET29-AtCCD1-His6 and micellar 8′-apo-β-caroten-8′-al with Triton X-100 was prepared according to (*7*). For the enzymatic cleavage 80 mL cell raw extract were mixed with 16 mL micellar substrate. The mixture was circulated in a standard pervaporation test module (GKSS, Geesthacht, Germany) at 24°C equipped with 35 cm² 2420G/V-1 membrane (GMT GmbH, Rheinfelden, Germany). The freeze traps were changed every hour and substrate and product concentrations determined by HPLC.

Figure 3 shows that 8′-apo-β-caroten-8′-al is cleaved within two hours by CCD1. After one hour the first permeate is collected and contains only little β-ionone, as the majority of the newly synthesized product is still situated in the retentate. Over the course of time the β-ionone changes places from the retentate to the permeate with the overall amount being constant as the substrate was depleted after two hours. The molar yield of the process was 0.6 with an enrichment factor of 3.

**Enzymatic Cleavage of β-Carotene**

Cell extract from *E.coli* pET29-AtCCD1-GST and micellar β-carotene was prepared according to (*1*). The mixture was circulated in a pervaporation module (GKSS, Geesthacht, Germany) equipped with 990 cm² POMS/PEI membrane (GKSS, Geesthacht, Germany). The freeze traps were changed at regular intervals and substrate and product concentrations determined by HPLC.

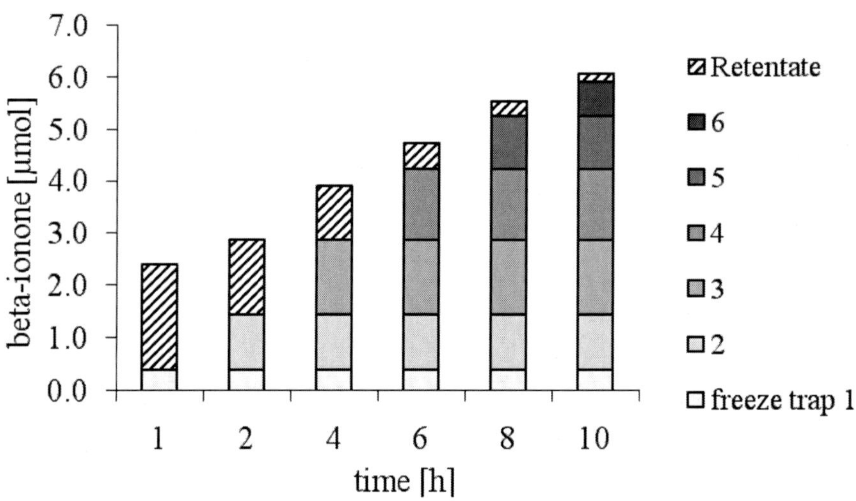


*Figure 4. β-ionone amounts in retentate and permeate during the enzymatic cleavage of β-carotene by AtCCD1 with in-situ product removal by organophilic pervaporation.*

In Figure 4 the substrate amount is not shown as it was present in excess. As in case of the enzymatic cleavage of 8′-apo-β-caroten-8′-al after one hour most of the product can be found in the retentate. In contrast to the first example the overall product concentration keeps increasing over ten hours albeit at a very low rate. Even with a high substrate concentration only little product is synthesized due to the very low reaction rate of the enzyme. The molar yield of the process was 0.38 with an enrichment factor of 21.

### Product Purity

For the pervaporation experiments the enzyme was not purified but cell raw extract was used. Figure 5 shows a GC-MS chromatogram of the permeate from the cleavage of β-carotene. Apart from the isomerization product α-ionone there are no side products detectable. Indole, which causes the unpleasant smell of *E. coli* cultures and would have been detrimental for the product quality, was not detected.

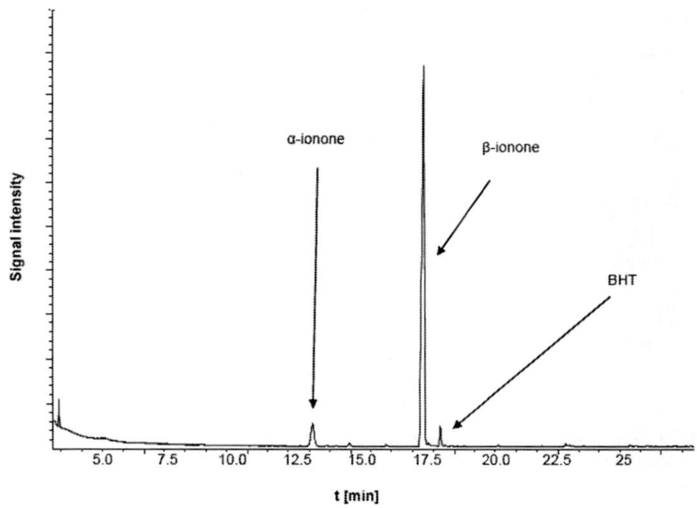


*Figure 5. GC-MS total ion chromatogram of permeate from the cleavage of β-carotene by AtCCD1. BHT (butylhydroxytoluene) is a solvent stabilizer.*

## Discussion and Outlook

A direct comparison of the yields and enrichment factors of the two examples is foregone as the experiments were done with different pervaporation membranes and in different experimental setups. However, in both cases our results show for the first time the successful coupling of an enzymatic cleavage with organophilic pervaporation as *in-situ* product removal.

For an unstable product such as β-ionone (*8*) organophilic pervaporation is very well suited. After its synthesis by carotenoid cleavage the product is transported through the pervaporation membrane. Upon its desorption from the membrane it evaporates and is instantly frozen and thus preserved by liquid nitrogen. The permeate is an aqueous-ethanolic solution of β-ionone making further downstream processing easy.

In comparison with other *in-situ* product removal techniques the advantages of organophilic pervaporation become particularly prominent. The transfer of the product into a second liquid, organic phase as well as the adsorption to porous materials (*9*) require an investment in the second phase material. Furthermore a solvent and consequently further investment for the desorption of the final product from the extractive phase is necessary. This kind of downstream processing is complicated in the European Union by a directive that only allows selected solvents for the production of food flavors (*10*). Organophilic pervaporation avoids these multiple phase transitions because the final product can usually be distilled right from the permeate.

In terms of product concentration there is room for optimization in the system presented. As Figure 2 shows, the membranes used are well suited for the enrichment of β-ionone. The cause for the low enrichment – even with efficient membranes – clearly needs to be sought in the limitations of the enzymatic system. The driving force for organophilic pervaporation is a high product concentration in the retentate. For the time being this is difficult to achieve because AtCCD1 is difficult to produce in large quantities in a heterologous host and at the same time the product β-ionone is unstable. If these shortcomings can be overcome organophilic pervaporation offers a huge potential for the technical application of carotenoid cleavage dioxygenases and other enzymatic flavor productions.

## References


1. Nacke, C. Ph.D. Thesis, Technische Universität München, Freising, Germany, 2010.
2. Schwartz, S. H.; Qin, X.; Zeevaart, J. A. *J. Biol. Chem.* **2001**, *276*, 25208–25211.
3. Jonquières, A.; Clément, R.; Lochon, P.; Néel, J.; Dresch, M.; Chrétien, B. *J. Membr. Sci.* **2002**, *206*, 87–117.
4. Etschmann, M. M.; Sell, D.; Schrader, J. *Biotechnol. Bioeng.* **2005**, *92*, 624–634.
5. Bluemke, W.; Schrader, J. *Biomol. Eng.* **2001**, *17*, 137–142.
6. Böddeker, K. W. *Pervaporation durch Membranen und ihre Anwendung zur Trennung von Flüssiggemischen*; Reihe 3: Verfahrenstechnik, Nr. 129; VDI: Düsseldorf, Germany, 1986.
7. Nacke, C.; Schrader, J. *J. Mol. Catal.: B* **2012**, *77*, 67–73.
8. Grivel, F.; Larroche, C.; Gros, J. B. *Biotechnol. Prog.* **1999**, *15*, 697–705.
9. Rodriguez-Bustamante, E.; Maldonado-Robledo, G.; Sánchez-Contreras, A.; Klimova, T.; Arreguin-Espinosa, R.; Sánchez, S. *Appl. Microbiol. Biotechnol.* **2006**, *71*, 568–573.
10. *Verordnung über die Verwendung von Extraktionslösungsmitteln und anderen technischen Hilfsstoffen bei der Herstellung von Lebensmitteln (Technische Hilfsstoff-Verordnung - THV)*, 1991; http://www.gesetzeim-internet.de/bundesrecht/elv/gesamt.pdf, last accessed 2012-07-19.

Chapter 16

# Oxidative Cleavage Products of Lycopene: Production and Reactivity in a Biomimetic Experimental Model of Oxidative Stress

**Michel Carail,[1,2] Pascale Goupy,[1,2] Eric Reynaud,[1,2] Olivier Dangles,[1,2] and Catherine Caris-Veyrat*,[1,2]**

**[1]INRA, UMR408 Sécurité et Qualité des Produits d'Origine Végétale, F-84000 Avignon, France**
**[2]Université d'Avignon et des Pays de Vaucluse, UMR408 Sécurité et Qualité des Produits d'Origine Végétale, F-84000 Avignon, France**
***E-mail: catherine.caris@avignon.inra.fr.**

Carotenoids oxidation products are of interest for their possible biological activity. We produced and characterized a series of lycopene oxidative cleavage products by oxidizing lycopene. We synthesized apo-lycopenoids of various polyene chain lengths and terminal functions. Antioxidant activity of apo-lycopenoids was determined using a lipid peroxidation test mimicking oxidative stress in the gastro-intestinal tract. Results allowed us to establish structure-activity relationships of antioxidant protection of apo-lycopenoids in the gastro-intestinal tract. A long carbon chain and a carboxylic acid endgroup conferring an amphiphilic character to the molecule are favorable to the antioxidant activity of apo-lycopenoids.

## Oxidative Cleavage Products of Carotenoids in Humans

Retinal and retinoic acid (i.e. vitamin A) are enzymatic oxidative cleavage products of provitamin A carotenoids that are essential for human health. The main biological functions of retinoids are related to vision and cellular differentiation. These products are produced from the central cleavage of provitamin A carotenoids. More recently, non-central oxidative metabolites of provitamin A carotenoids and oxidative metabolites of non-provitamin A carotenoids have been suggested to exert biological effects in humans, either in the prevention (*1*, *2*) or the promotion of degenerative diseases, e.g. in the ATBC (*3*) or CARET studies (*4*). It has been proposed that these carotenoid oxidation products may serve as precursors of retinoids, which would be the active species mediating the biological effects of vitamin A, or to their own specific activities which have yet to be identified (*5*). In this latter case, for example, oxidative cleavage products of lycopene could be biologically active. Although a number of biological studies have been performed *in vitro* for oxidative cleavage of non provitamin A carotenoids (*6*), *in vivo* effects in humans are not known. It is only very recently that non provitamin A carotenoid oxidative cleavage products have been found for the first time *in vivo* in humans. These products are derivatives of lycopene, and are collectively refered to as apo-lycopenoids. Kopec et al. (*7*) found several apo-lycopenals in plasma of volunteers who had ingested tomato purée. Since apo-lycopenals were also found in the tomato product, the molecules could not be qualified as lycopene metabolites formed *in vivo*.

Herein, we present methods for the production of apo-lycopenoids by two different strategies: oxidation of lycopene and organic synthesis. Finally, we will present studies on their antioxidant activity in an *in vitro* system mimicking oxidative stress in the gastro-intestinal tract.

## Chemical Oxidation of Lycopene

The work presented here was designed to characterize oxidation products of lycopene obtained by two different systems of oxidation: potassium permanganate and oxygen with a metalloporphyrin as a catalyst.

Because of the hydrophilic character of potassium permanganate and the lipophilicity of lycopene, the oxidation was performed in a biphasic medium using cetyltrimethylammonium bromide as phase transfer catalyst at room temperature. Samples were taken periodically, and the reaction was followed by high performance liquid chromatography-diode array detection-mass spectrometry (HPLC-DAD-MS) (*8*). After 1 h of incubation, lycopene was completely consumed and a range of products was formed. Compounds were separated on a C30 HPLC column, and almost all products were tentatively identified by comparing their molecular mass and their UV-visible spectra data with literature when available. Two types of oxidation compounds were obtained: those resulting from a single oxidative cleavage and those derived from a double oxidative cleavage. Mono-oxidative cleavage compounds produced included apo-lycopenals (posessing an aldehyde endgroup) and apo-lycopenones (posessing a ketone endgroup). Apo-lycopenals/ones of all possible chain lengths were detected (Figure 1). For the first time, apo-11-lycopenal was purified by preparative HPLC and fully characterized by $^{1}H$ and $^{13}C$ NMR. In addition, double oxidative cleavage compounds, diapo-carotendials, were detected (Figure 2), among which diapo-8,6′-lycopendial was purified and fully characterized for the first time.

apo-6'-lycopenal (6'-CHO)

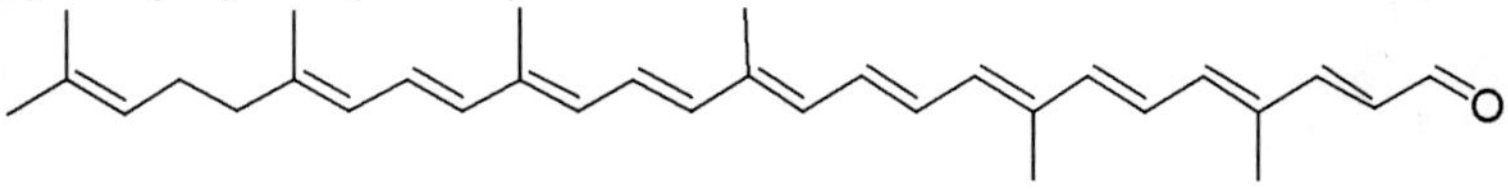

apo-8'- lycopenal (8'-CHO)

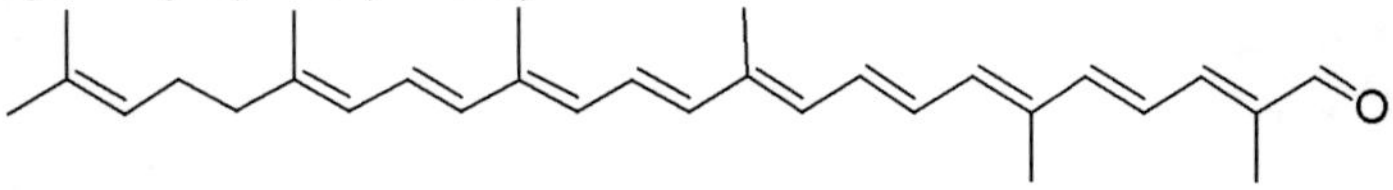

apo-10'-lycopenal (10'-CHO)

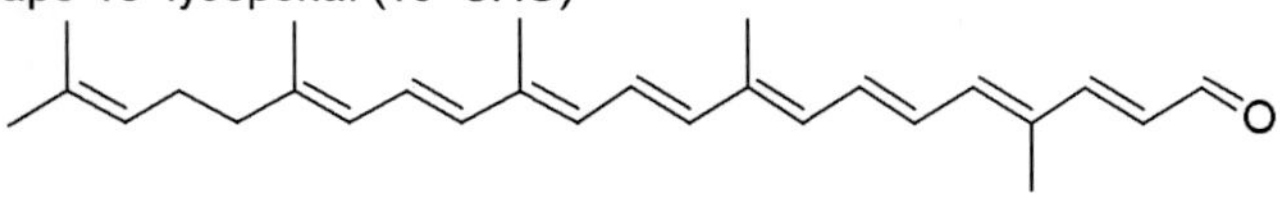

apo-12'-lycopenal (12'-CHO)

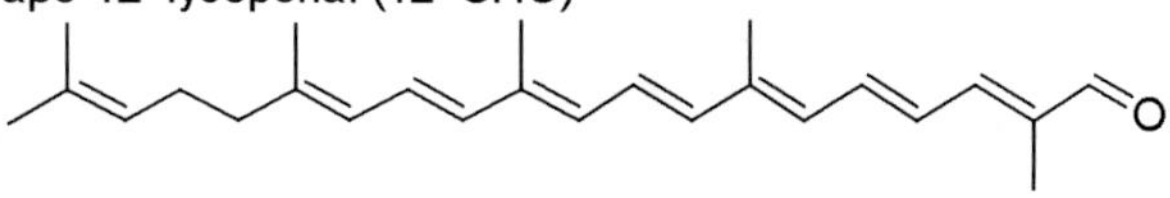

apo-14'-lycopenal (14'-CHO)

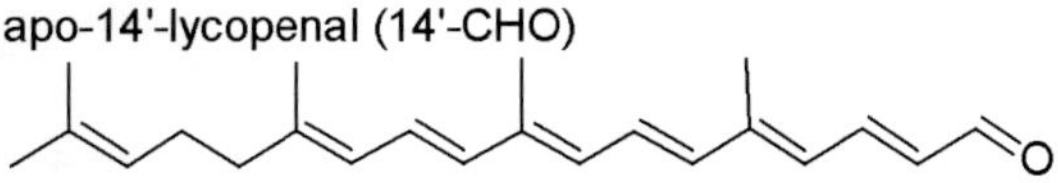

apo-15-lycopenal = acycloretinal

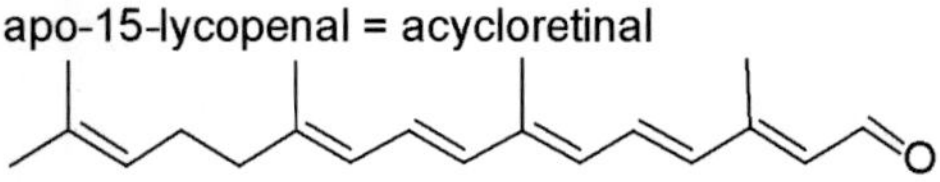

apo-13-lycopenone = 6,10,14-trimethyl-3,5,7,9,13-pentadecapentaene-2-one

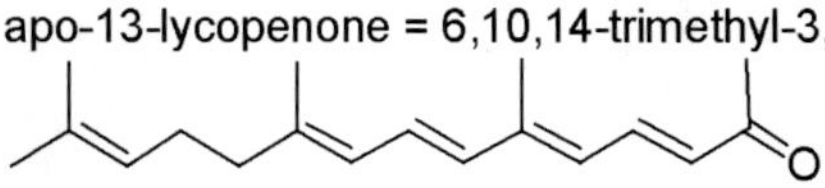

apo-11-lycopenal = 3,7,11-trimethyl-2,4,6,10-dodecatetraene-1-al (11-CHO)

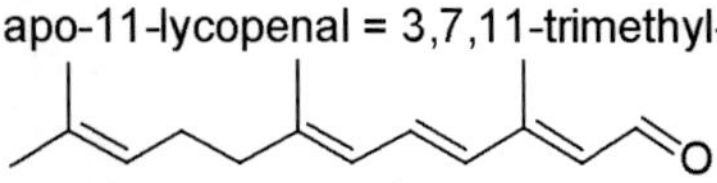

apo-9-lycopenone = 6,10-dimethyl-3,5,9-undecatriene-2-one

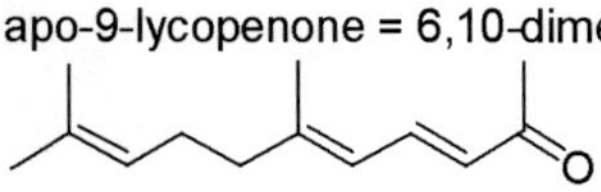

apo-7-lycopenal = 3,7-dimethyl-2,6-octadiene-1-al

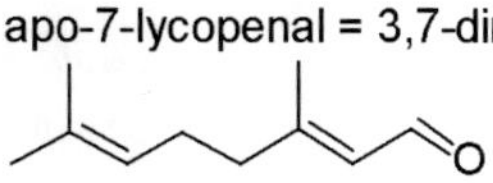

apo-5-lycopenone = 2-methyl-2-heptene-6-one

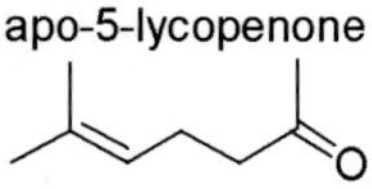

*Figure 1. Apo-lycopenals and apo-lycopenones obtained by oxidation of lycopene with potassium permanganate in a biphasic medium.*

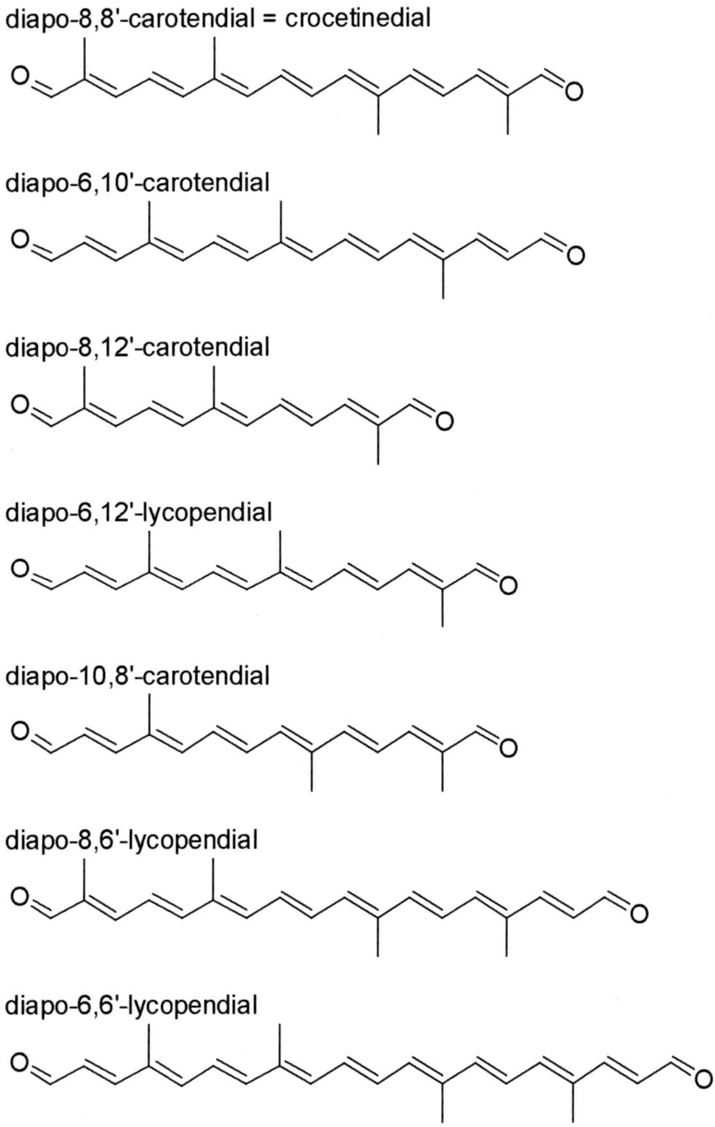


*Figure 2. Diapo-carotendials and diapo-lycopendials obtained by oxidation of lycopene with potassium permanganate in a biphasic medium.*

The catalytic oxidation activity of carbonyl ruthenium(II)-tetraphenylporphyrin towards lycopene was also studied (*8*). We have previously tested a similar system with β-carotene (*9*). Metalloporphyrins mimic the active center of cytochrome P450 enzymes, which are involved in the *in vivo* transformation of xenobiotics. The reaction of carbonyl ruthenium(II)-tetraphenylporphyrin with lycopene was slow enough to be followed by HPLC-DAD-MS over a period of 96 h. Even after the complete disappearance of lycopene, the composition of the reaction mixture changed, indicating that the catalytic system was capable of further oxidizing products formed initially in the reaction. Three types of products were detected: (*Z*)-isomers of lycopene, lycopene epoxides, and apo-lycopenals/ones. Almost all of the apo-lycopenals/ones previously identified after the oxidation of lycopene with potassium permanganate were detected, except the short-chain apo-5-lycopenone and apo-7-lycopenal. We anticipate that these short-chain products are volatile and thus they likely evaporated during the course of the experiment.

Semi-quantitative data on the accumulation and consumption of these different lycopene derived-products throughout the reaction suggests a possible mechanism of formation (Figure 3). (*Z*)-Isomers of lycopene are the first products to be formed. These isomers are then oxidized to various lycopene epoxides. Next, these epoxide groups are oxidatively cleaved to produce apo-lycopenals/ones. Finally, the long-chain apo-lycopenals (apo-15- to apo-6′-lycopenal) would be oxidatively cleaved and transformed into shorter apo-lycopenals/ones (apo-9- to apo-13-lycopenal/ones).

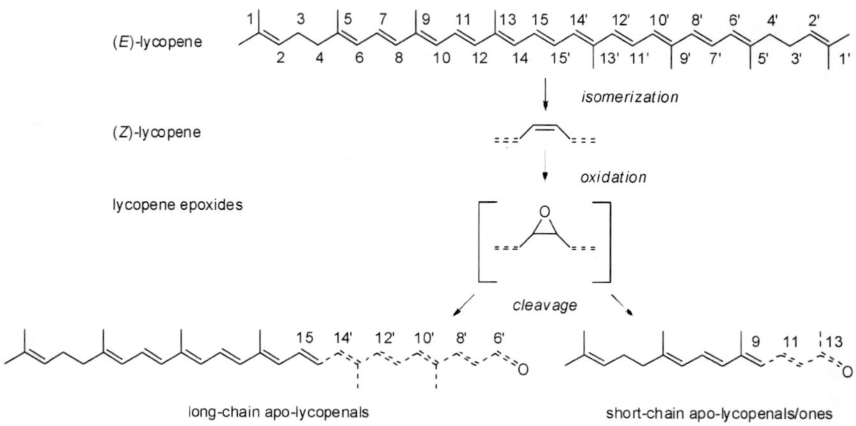


*Figure 3. Possible mechanisms of oxidation of lycopene by molecular oxygen catalyzed by a metalloporphyrin. (adapted from (8)).*

Oxidation of carotenoids has previously been performed using different reagents: potassium permanganate, ozone, chromic acid, manganese dioxide etc (*10*). In addition, some oxidation products of lycopene have also been obtained using a combination of osmium tetroxide and hydrogen peroxide (*11*,

12). The methods we developed and used with lycopene proved to be efficient routes of obtaining all the possible apo-lycopenoids described above, which were tentatively identified by HPLC-DAD-MS. These methods even allowed us to obtain some products in high enough yield to confirm identity by NMR. *In vivo*, it is believed that carotenoid metabolism is mainly catalyzed by carotenoid cleavage enzymes (β-carotene oxygenase 1 and 2, i.e. BCO1 and BCO2) with an active center containing a ferrous iron atom (*13*) coordinated by histidine residues (*14*, *15*). In addition, some evidence in rodents suggested that lycopene may be metabolized by specific cytochrome P450 enzymes (*16*). Thus, the *in vitro* oxidation catalyzed by metalloporphyrin enables insights into the possible *in vivo* mechanisms of lycopene cleavage.

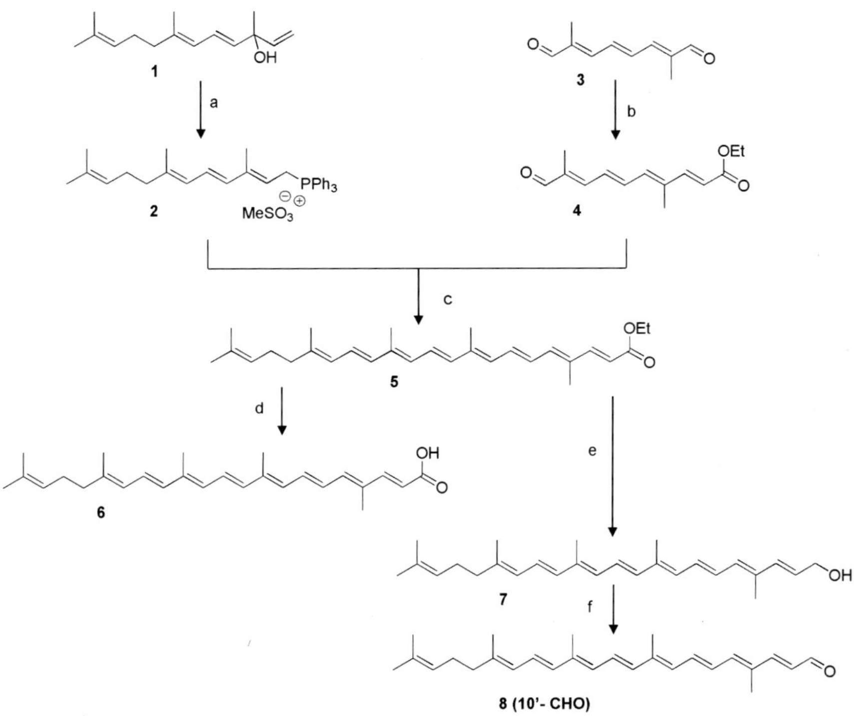


*Figure 4. Chemical synthesis of apo-10'-lycopenoids. Reagents and conditions: (a) $PPh_3$, $MeSO_3H$, AcOH, 10°C→45°C, 71%; (b) $(EtO)_2OPCH_2COOEt$, tBuOK, THF, 38%; (c) NaOMe, $CH_2Cl_2$, 18%; (d) NaOH, EtOH, 55%, (e) DIBAH, THF, 31%; (f) $MnO_2$, AcOEt, 44%.*

## Synthesis of Apo-Lycopenoids

In order to obtain apo-lycopenoids in a sufficient amount to perform physico-chemical and antioxidant studies, we developed a synthetic strategy based on previous work describing the total synthesis of retinoids and carotenoids (*17*) of (all-*E*)-lycopene (*18*) and of (*Z*)-isomers of lycopene and apo-12′-lycopenal (*19*). To build the hydrocarbon skeleton of the apo-10′ and apo-14′-lycopenoids, we used chemical reactions classically utilized in carotenoid synthesis, such as the Horner-Wadsworth-Emmons (HWE) and Wittig condensation reactions. Our aim was to obtain apo-lycopenoids in a pure (*E*) configuration. Thus, a critical point in the synthesis was the diastereoisomeric control of the carbon-carbon double bond formed. Triphenylphosphonium methanesulfonate salt is a common starting material for the synthesis of the targeted apo-10′ and apo-14′-lycopenoids (Figures 4 and 5). We prepared this phosphonium salt from vinylpseudoionol **1** using methanesulfonic acid in glacial acetic acid (*18*). **2** was obtained in good yield (71%) and with acceptable stereoselectivity (*E/Z* = 4). The other primary structure necessary for synthesis of apo-10′-lycopenoids, **4**, was prepared by elongation of dialdehyde **3** in a HWE reaction, with 1 equivalent of the triethylphosphonoacetate giving **4** with a high diastereoisomeric (*E/Z*) ratio of 25 and a yield of 38% after purification by open column chromatography and subsequent crystallization. A Wittig condensation between **2** and **4** gave apo-10′-lycopenoic ethyl ester **5** in pure (*E*) form with a yield of 18% upon crystallization. Saponification of **5** with sodium hydroxide in ethanol gave apo-10′-lycopenoic acid **6** in 55% yield. Reduction of **5** with DIBAH in THF provided apo-10′-lycopenol 7 in 31 % yield, which was converted into apo-10′-lycopenal **8** with a 44% yield upon subsequent oxidation using $MnO_2$ in ethyl acetate.

For the synthesis of the apo-14′-lycopenoids, product **13**, was obtained in a four-step pathway from ethyl 3-methyl-4-oxocrotonate **9** (*20*): (i) protection of the aldehyde group of **9**, (ii) reduction of the ethyl ester into the corresponding allylic alcohol, (iii) oxidation into aldehyde **11**, and (iv) HWE condensation in order to elongate the carbon chain by 2 carbons. After deprotection of the aldehyde function, the target product **13** was obtained with a global yield of 12% after crystallization. A Wittig condensation between **2** and **13,** as described previously for compound **5**, provided apo-14′-lycopenoic methyl ester **14** in 24% yield upon crystallization; ethyl ester of **13** was converted into a methyl ester during the reaction. Apo-14′-lycopenoic methyl ester **14** was then transformed either into apo-14′-lycopenoic acid **15** by saponification with a yield of 49%, or into apo-14′-lycopenol **16** (yield 35%) which in turn gave apo-14′-lycopenal **17** in 44% yield upon oxidation and purification.

A similar synthetic strategy, based on the HWE reaction, was developed to generate the hydrocarbon skeleton of apo-11-lycopenoids (Figure 6). Apo-11-lycopenoic ethyl ester **19** was obtained by elongation of the carbon chain of (*E*) pseudo-ionone **18** upon treatment with one equivalent of triethylphosphonoacetate. The reaction led to **19** (82% yield) with the new carbon–carbon double bond exhibiting a diastereoisomeric (*E/Z*) ratio of 3.5. The crude mixture was reduced with DIBAH to give apo-11-lycopenol **20** with a yield of 16% after purification, with 97% as the (*E*) isomer. Oxidation of the allylic alcohol of **20** provided apo-11-lycopenal **21** in 30% yield after purification (97% (*E*) isomer). Further mild oxidation of the aldehyde function of **21** using $Ag_2O$ gave apo-11-lycopenoic acid **22** as a pure (*E*) isomer with a yield of 90%.

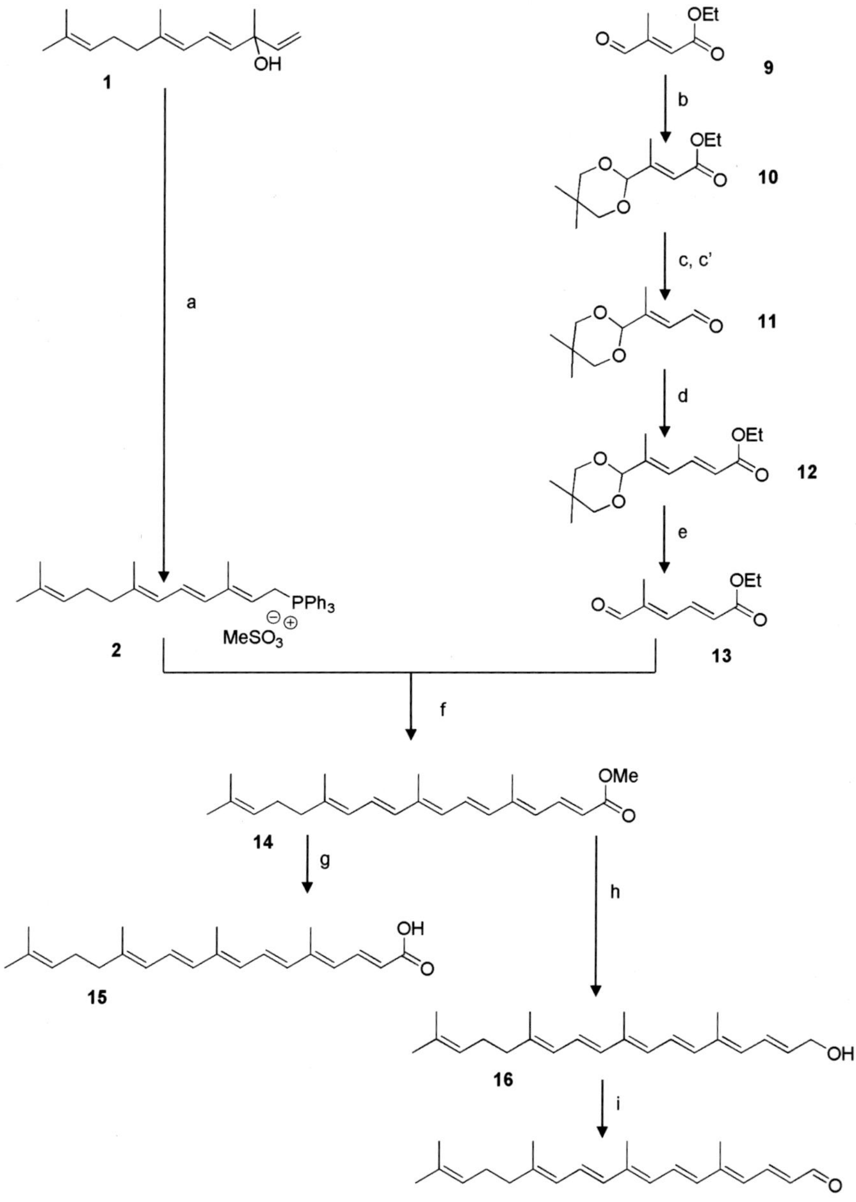


*Figure 5. Chemical synthesis of apo-14'-lycopenoids. Reagents and conditions: (a) $PPh_3$, $MeSO_3H$, AcOH, 10°C→45°C, 71%; (b) 2,2-dimethylpropanediol, APTS, toluene, reflux, 81%; (c) DIBAH, THF, (c') $MnO_2$, EtOAc 56%; (d) $(EtO)_2OPCH_2COOEt$, tBuOK, THF, 55%; (e) HCl, $MeOH/H_2O$, 46%; (f) NaOMe, $CH_2Cl_2$, 24%; (g) NaOH, EtOH, 49%, (h) DIBAH, THF, 35%; (i) $MnO_2$, AcOEt, 44%.*

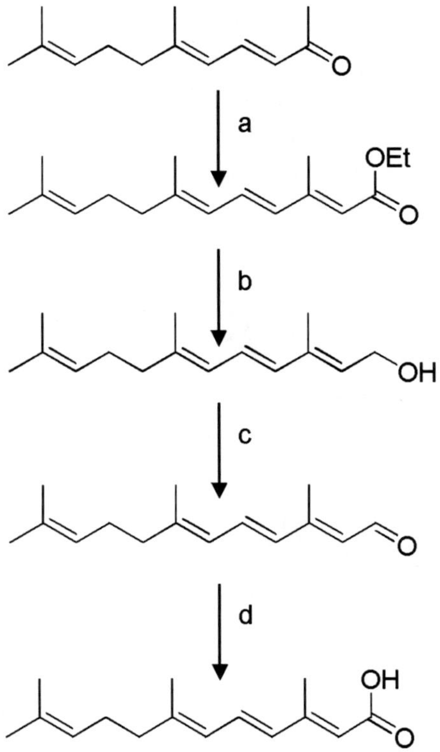


18

19

20

21 (11-CHO)

22

*Figure 6. Chemical synthesis of apo-11-lycopenoids. Reagents and conditions: (a) triethylphosphonoacetate, NaH, THF, 82%, (b) DIBAH, THF, 16%, (c) $MnO_2$, AcOEt, 30%, (d) $Ag_2O$, NaOH, EtOH, 90%.*

Apo-lycopenoids of three different chain lengths (apo-11, apo-14′- and apo-10′) and different functional endgroups (ester, carboxylic acid, alcohol and aldehyde) were thus obtained for the first time by organic synthesis in the (*E*) stereoisomeric form.

## Antioxidant Activity of Apo-Lycopenoids in a Biomimetic Experimental Model of Oxidative Stress

Carotenoids can serve as antioxidants *in vitro*. However, they are poorly bioavailable, and their concentrations in plasma (on the order of 1 μM) seem quite low for carotenoids to serve as a significant and direct source of antioxidants in circulation (*21*). In contrast, in humans it has been demonstrated that lycopene concentrations in the stomach reach 50 μM only 20 min after ingestion of tomato purée (containing 10 mg of lycopene). At 1 hr and 2 hr after tomato purée consumption, lycopene levels still remain quite high (at 40 μM and 25 μM, respectively) (*22*). Thus, the antioxidant activity of poorly bioavailable phytochemicals such as carotenoids and polyphenols may have relevance in the gastro-intestinal tract during digestion (*23*, *24*).

The stomach is a bioreactor in which oxidative conditions can occur after ingestion of a meal, leading to the production of potentially toxic lipid peroxides. These peroxides are considered a possible risk factor for the development of atherogenesis (*25*). Oxidation of ingested fatty acids can be catalyzed by iron, which is more redox active due to the acidic pH of the stomach. Phytochemicals ingested from fruits and vegetables could participate in the inhibition of this lipid peroxidation.

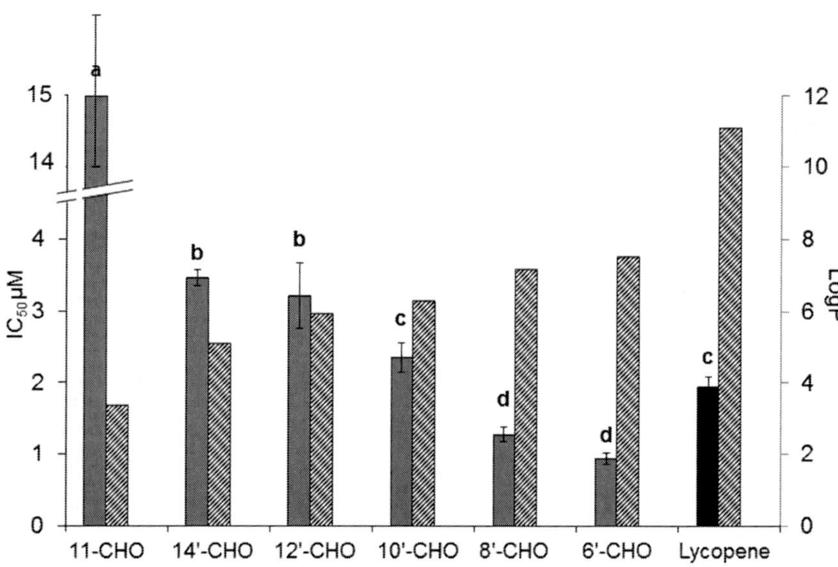


*Figure 7. Antioxidant ($IC_{50}$) activity (solid grey and black histograms) and lipophilicity (LogP values, hatched histograms) of apo-lycopenals and lycopene. Histogram for apo-11-lycopenal was truncated because of the very high $IC_{50}$ value.*

We have designed a chemical experimental model to evaluate the antioxidant activity of phytochemicals, including carotenoids and their metabolites, using simple conditions for mimicking the stomach (*26*). The model consists of a mildly acidic (pH 5.8) micellar solution in which oxidative stress is generated by metmyoglobin, one of the main forms of dietary iron found in the gastric compartment after the consumtion of red meat. Linoleic acid is added to mimic the unsaturated fatty acids ingested from a meal. Linoleic acid peroxidation is catalysed by metmyoglobin, and lipid peroxides are monitored by UV-visible spectrophotometry at 234 nm, the characteristic wavelength of the conjugated diene function in the molecule. The presence of antioxidant molecules, like carotenoids or polyphenols, inhibits the appearance of lipid peroxides. We have previously shown that hydrophilic and lipophilic antioxidants display distinct inhibition mechanisms (*27*). We previously used this experimental model to discriminate the antioxidant activity of red wine anthocyanins derivatives with similar structures (*28*).

In the present section, we present information on the ability of (*E*)-lycopene and apo-lycopenoids with different chain lengths and terminal groups (ester, carboxylic acid, alcohol and aldehyde) to inhibit lipid peroxidation in the simple chemical model of postprandial oxidative stress in the gastric compartment (*29*).

The accumulation of the primary products of lipid oxidation, *i.e.* conjugated dienes, was monitored by UV-visible spectroscopy at 234 nm in the absence (uninhibited peroxidation) or presence of (*E*)-lycopene or apo-lycopenoids at different concentrations. The half maximal inhibitory concentration ($IC_{50}$) parameter was calculated for a global estimation of the antioxidant efficiency (*29*). The lower the $IC_{50}$ value, the more potent the antioxidant. We compared the activity of six apo-lycopenals (3 synthesized and 3 commercially available) with those of lycopene (Figure 7). Apo-6′ and apo-8′-lycopenals are more efficient antioxidants than lycopene. Apo-10′-lycopenal is a less potent antioxidant than apo-8′-lycopenal, followed by apo-12′ and apo-14′-lycopenals. The high $IC_{50}$ value of apo-11-lycopenal denotes its very low antioxidant activity. Thus, antioxidant activity of apo-lycopenals appears to increase with chain length, from the shortest apo-11-lycopenal to the longest apo-6′-lycopenal. The same trend was observed for apo-lycopenoic acids, apo-lycopenoic esters, and apo-lycopenols (results not shown, (*29*)). Antioxidant activity was positively correlated with the lipophilicity of the apo-lycopenals. The higher the LogP (calculated using the ChemPro program, ChemDraw Pro11.0 CambridgeSoft), the more lipophilic the compound, and the more antioxidant potency observed.

As shown for the apo-10′ and apo-14′-lycopenoids (Figures 8 and 9), for a fixed chain length, the terminal function influences the antioxidant activity in a similar manner: the apo-lycopenoic acids are the most efficient antioxidants followed by apo-lycopenals > apo-lycopenoic esters > apo-lycopenols. Because of their anionic carboxylate end, apo-lycopenoic acids are amphiphilic antioxidants which likely locate at the interface of micelles. We hypothesize that this position improves their efficiency at inhibiting the peroxidation mechanism.

Apo-10'-lycopenoids

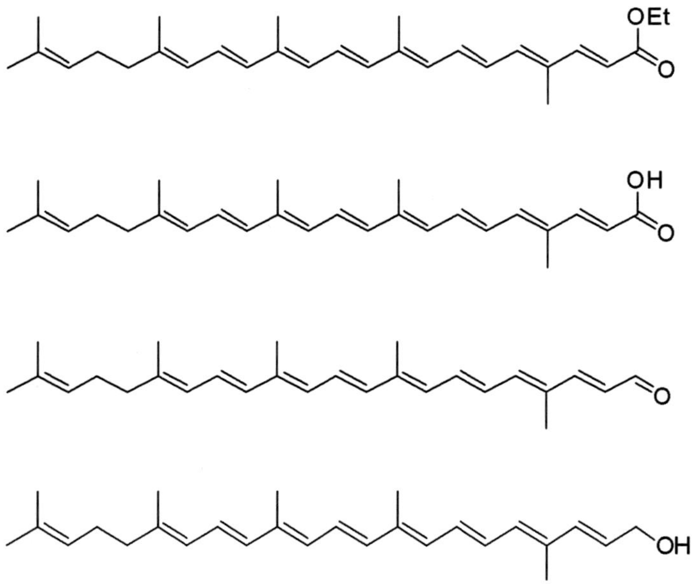


Apo-14'-lycopenoids

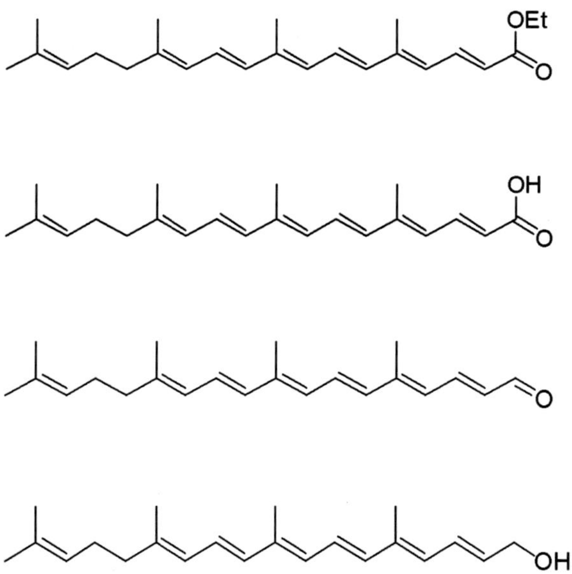


*Figure 8. Synthetic apo-10'-lycopenoids and apo-14'-lycopenoids with different ending functions.*

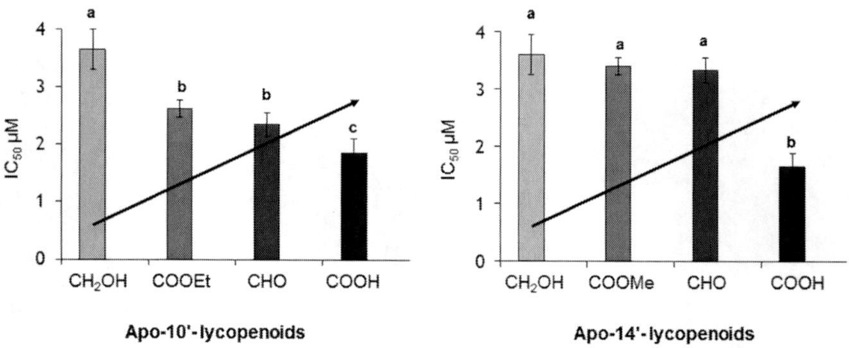


*Figure 9. Influence of the terminal group on the antioxidant activity of apo-lycopenoids. Arrows indicate increasing antioxidant activity.*

There is more interest in oxidative cleavage products of lycopene since they have been found *in vivo* in humans (*7*) ferrets (*30*) and rats (*31*). While their concentration in human plasma (in the nM range) seems too low for them to exert an antioxidant effect in circulation, the levels of these apo-lycopenoids in other compartments (for example, the gastro-intestinal tract) are simply not known. Further work is needed to detect and quantify apo-lycopenoids in various compartments *in vivo* to understand the role(s) they may play in exerting antioxidant or other effects. Moreover, even though apo-lycopenoids have been found in human plasma, it is not yet proven that they are lycopene metabolites formed *in vivo*. In addition, these compounds may exert biological effects via other mechanisms besides antioxidant activity. Cellular signaling activities *in vitro* have been demonstrated for various apo-lycopenoids. For example, apo-10′-lycopenoic acid was shown to inhibit lung cancer cell growth *in vitro* and suppresses lung tumorigenesis in the A/J mouse model *in vivo* (*32*), to induce Nrf2-mediated expression of phase II detoxifying/ antioxidant enzymes in human bronchial epithelial cells (*33*), and to impact adipose tissue biology *via* the retinoic acid receptors (*34*). Diapocaroten-dials were also shown to stimulate the electrophile/antioxidant response element (EpRE/ARE) *in vitro* and thus could serve as active mediators in the stimulation of the EpRE/ARE system by these products *in vivo* (*35*).

In conclusion, we have generated numerous apo-lycopenoids via multiple methods of lycopene oxidation and chemical synthesis. In addition, we have created a novel *in vitro* system to mimic lycopene oxidation by metalloporphyrins, and proposed the mechanism by which these compounds may be made. Finally, we tested the ability of various apo-lycopenoids to protect lipids from oxidation during digestion in the gastric compartment. Further research studies are needed to better understand how lycopene is metabolized *in vivo*. In addition, future work should focus on the *in vivo* biological activity of lycopene oxidative cleavage compounds.

## Acknowledgments

Authors thank Hansgeorg Ernst from BASF (Ludwigshafen, Germany), Werner Simon and Bettina Wüstenberg from DSM Nutritional Product (Kaiseraugst, Switzerland) for kindly providing starting materials for the organic synthesis work. They also thank Rachel Kopec for her help with the manuscript.

## References


1. Khachik, F.; Beecher, G.; Smith, J. C. *J. Cell. Biochem.* **1995**, *22*, 236–246.
2. King, T. J.; Khachik, F.; Bortkewicz, H.; Fukushima, S.; Morioka, S.; Bertram, J. S. *Pure Appl. Chem.* **1997**, *69*, 2135–2140.
3. The alpha-Tocopherol beta-Carotene Cancer Prevention Study Group. *N. Eng. J. Med.* **1994**, *330*, 1029–1035.
4. Omenn, G. S.; Goodman, G. E.; Thornquist, M. D.; Balmes, J.; Cullen, M. R.; Glass, A.; Keogh, J. P.; Meyskens, F. L., Jr; Valanis, B.; Williams, J. H., Jr; Barnhart, S.; Hammar, S. *N. Eng. J. Med.* **1996**, *334*, 1150–1155.
5. Wang, X. D. In *Carotenoids in Health and Disease*; Krinsky, N. I., Mayne, S. T., Sies, H., Eds.; Marcel Dekker: New York, 2004; pp 313−335.
6. Caris-Veyrat, C. In *Food Colorants - Chemical and Functional Properties*; Socaciu, C., Ed.; CRC Press: Boca Raton, FL, 2008; pp 177−192.
7. Kopec, R. E.; Riedl, K. M.; Harrison, E. H.; Curley, R. W.; Hruszkewycz, D. P.; Clinton, S. K.; Schwartz, S. J. *J. Agric. Food Chem.* **2010**, *58*, 3290–3296.
8. Caris-Veyrat, C.; Schmid, A.; Carail, M.; Böhm, V. *J. Agric. Food Chem.* **2003**, *51*, 7318–7325.
9. Caris-Veyrat, C.; Amiot, M. J.; Ramasseul, R.; Marchon, J. C. *New J. Chem.* **2001**, *25*, 203–206.
10. Caris-Veyrat, C. In *Carotenoids - Physical, Chemical, and Biological Functions and Properties*; Landrum, J., Ed.; CRC Press: Boca Raton, FL, 2010; pp 215−228.
11. Aust, O.; Ale-Agha, N.; Zhang, L.; Wollersen, H.; Sies, H.; Stahl, W. *Food Chem. Toxicol.* **2003**, *41*, 1399–1407.
12. Wingerath, T.; Kirsch, D.; Spengler, B.; Stahl, W. *Anal. Biochem.* **1999**, *272*, 232–242.
13. von Lintig, J. In *Colors with Functions: Elucidating the Biochemical and Molecular Basis of Carotenoid Metabolism*; Cousins, R. J., Ed.; 2010; Vol. 30, pp 35−56.
14. Kloer, D. P.; Ruch, S.; Al-Babili, S.; Beyer, P.; Schulz, G. E. *Science* **2005**, *308*, 267–269.
15. Poliakov, E.; Gentleman, S.; Cunningham, F. X.; Miller-Ihli, N. J.; Redmond, T. M. *J. Biol. Chem.* **2005**, *280*, 29217–29223.
16. Breinholt, V.; Lauridsen, S. T.; Daneshvar, B.; Jakobsen, J. *Cancer Lett.* **2000**, *154*, 201–210.
17. Paust, J. *Pure Appl. Chem.* **1991**, *63*, 45–8.
18. Ernst, H. *Pure Appl. Chem.* **2002**, *74*, 2213–2226.
19. Hengartner, U.; Bernhard, K.; Meyer, K.; Englert, G.; Glinz, E. *Helv. Chim. Acta* **1992**, *75*, 1848–1865.

20. Reynaud, E.; Aydemir, G.; Ruehl, R.; Dangles, O.; Caris-Veyrat, C. *J. Agric. Food Chem.* **2011**, *59*, 1457–1463.
21. Lowe, G. M.; Bilton, R. F.; Davies, I. G.; Ford, T. C.; Billington, D.; Young, A. J. *Ann. Clin. Biochem.* **1999**, *36*, 323–332.
22. Tyssandier, V.; Reboul, E.; Dumas, J. F.; Bouteloup-Demange, C.; Armand, M.; Marcand, J.; Sallas, M.; Borel, P. *Am. J. Physiol.: Gastrointest. Liver Physiol.* **2003**, *284*, G913–G923.
23. Kanner, J.; Lapidot, T. *Free Radical Biol. Med.* **2001**, *31*, 1388–1395.
24. Halliwell, B.; Zhao, K.; Whiteman, M. *Free Radical Res.* **2001**, *33*, 819–830.
25. Cohn, J. S. *Curr. Opin. Lipidol.* **2002**, *13*, 19–24.
26. Vulcain, E.; Goupy, P.; Caris-Veyrat, C.; Dangles, O. *Free Radical Res.* **2005**, *39*, 547–563.
27. Goupy, P.; Vulcain, E.; Caris-Veyrat, C.; Dangles, O. *Free Radical Biol. Med.* **2007**, *43*, 933–946.
28. Goupy, P.; Bautista-Ortin, A. B.; Fulcrand, H.; Dangles, O. *J. Agric. Food Chem.* **2009**, *57*, 5762–5770.
29. Goupy, P.; Reynaud, E.; Dangles, O.; Caris-Veyrat, C. *New J. Chem.* **2012**, *36*, 575–587.
30. Hu, K. Q.; Liu, C.; Ernst, H.; Krinsky, N. I.; Russell, R. M.; Wang, X. D. *J. Biol. Chem.* **2006**, *281*, 19327–19338.
31. Gajic, M.; Zaripheh, S.; Sun, F. R.; Erdman, J. W. *J. Nutr.* **2006**, *136*, 1552–1557.
32. Lian, F.; Smith, D. E.; Ernst, H.; Russell, R. M.; Wang, X. D. *Carcinogenesis* **2007**, *28*, 1567–1574.
33. Lian, F.; Wang, X. D. *Int. J. Cancer* **2008**, *123*, 1262–1268.
34. Gouranton, E.; Aydemir, G.; Reynaud, E.; Marcotorchino, J.; Malezet, C.; Caris-Veyrat, C.; Blomhoff, R.; Landrier, J. F.; Ruehl, R. *Biochim. Biophys. Acta* **2011**, *1811*, 1105–1114.
35. Linnewiel, K.; Ernst, H.; Caris-Veyrat, C.; Ben-Dor, A.; Kampf, A.; Salman, A.; Danilenko, M.; Levy, J.; Sharoni, Y. *Free Radical Biol. Med.* **2009**, *47*, 659–667.

## Chapter 17

# Intramolecular Cyclisation as Structural Transformation of Carotenoids During Processing of Paprika (*Capsicum annuum* L.) and Paprika Oleoresins

**Elisabet Fernández-García, José J. Ríos, Manuel Jarén-Galán, and Antonio Pérez-Gálvez***

**Group of Chemistry and Biochemistry of Pigments, Food Biotechnology Department, Instituto de la Grasa (CSIC), Av. Padre García Tejero 4, 41012, Sevilla, Spain**
***E-mail: aperez@cica.es.**

Intramolecular cyclisation stands out as the main reaction of structural transformation of carotenoids in paprika and paprika oleoresins. This process as well as those that subsequently take place, i.e. elimination of an in-chain unit or heterolytic fragmentation, are promoted by thermal processing, either during the dehydration of the fruits, the extraction of the lipophilic content with organic solvent or at the evaporation step to remove organic solvent and its residues. The heat-induced degradation yields a degradation profile formed by *nor*-carotenoids, that is, products with end-groups identical to those in the parent compound, but with a shorter polyene skeleton. This reaction process contributes to the aroma profile with specific compounds like toluene and *m*-xylene which arise by simple cyclisation and elimination reactions of the polyene.

## Introduction

Since the discovery of the New World, red pepper fruits spread widely to the rest of the world. Today, this fruit is grown for commercial purposes in the United States, Brazil, India, Taiwan, South Africa, Zimbabwe, and throughout Europe, with Hungary and Spain as the most representative producers of *capsicum* fruits. Pepper plant (*capsicum* genus) belongs to the Solanaceae family like potato, tomato, tobacco, and petunia. *Capsicum* genus includes 22 wild species and five domesticated species: *C. annuum*, *C. baccatum*, *C. chinese*, *C. frutescens*, and *C. pubescens*, with *C. annuum* and *C. frutescens* being the most economically important. Fruits of these plants are used in the manufacture of selected commercial products known for their color and pungency respectively, mainly paprika and paprika oleoresins. *Capsicum annuum* L. plant is extensively cultivated, producing fruits that vary in length, color, and pungency depending upon the cultivar. The nonpungent cultivars are used for direct consumption, either fresh or dried, whole or ground. They are also widely used as natural colorants to improve or provide that organoleptic attribute in other foodstuffs.

*Capsicum* fruits and related products may be used to improve the presentation (based on the coloring capacity owed to carotenoids) or to deliver a characteristic flavor (based on the pungency owed to capsaicinoids). To facilitate transport and to enhance handling during the storage and dosage of spices in food formulations, it is usual to obtain the paprika, the dry ground powder, or paprika oleoresin, the oily concentrate that contains the carotenoid pigments, in the case of *capsicum*, the carotenoid pigments and the capsaicinoids. Two types of *capsicum* oleoresins are distinguished depending on their use: for coloring purposes, paprika oleoresin (10-15% of color content), and for pungency and color uses, *capsicum* oleoresin (3-6% of capsaicinoid content, 1% of color content).

## Carotenoids as Components Responsible for Fruit Color

The fruits of C. *annuum* owe their intense red color to carotenoid pigments that are synthesized massively during fruit ripening. Carotenoids are isoprenoid compounds with an extensive system of conjugated double bonds (polyene chain) which is responsible for their physical and chemical properties. This characteristic structure is the primary cause for their lipophilic behaviour, as well as for their distinctive light-absorbing properties and therefore of their coloring capacity. All the carotenoid pigments present in the pepper contain nine conjugated double bonds in the central polyene chain. The presence of different end groups (β, ε, κ, 3-hydroxy-5,6-epoxide), changes the chromophore properties of each pigment, allowing them to be classified in two isochromic families: red and yellow (Figure 1). The red fraction contains capsanthin, 5,6-epoxide-capsanthin, and capsorubin, which are almost exclusive to the *capsicum* genus (*1–3*), while the yellow fraction comprises the rest of the pigments (β-carotene, β-cryptoxanthin, zeaxanthin, antheraxanthin, violaxanthin, and cucurbitaxanthin A). Ratio between the two fractions should remain invariable during processing, but it has been shown that heat treatment results in a decrease in the total pigment concentration, with the initial proportion of red and yellow pigments switching in favor of one or the other depending on the temperature used. At temperatures below 60 °C, the rate of destruction of the yellow pigment fraction was higher than that of the red pigments. Above 60 °C, the order of lability of the two fractions was inverted, and the red pigment fraction became the more unstable. Thus, treatments above this temperature increase the relative proportion of yellow pigments (*4*, *5*).

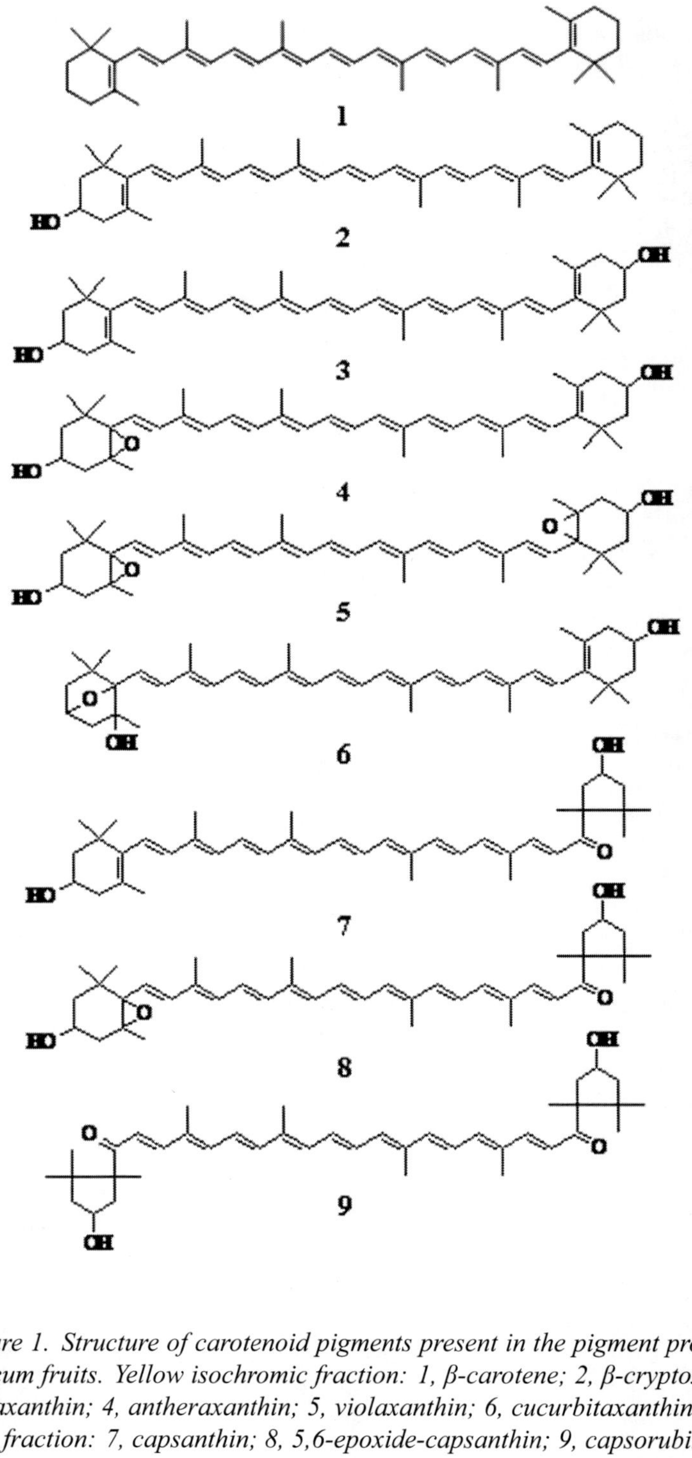


*Figure 1. Structure of carotenoid pigments present in the pigment profile of capsicum fruits. Yellow isochromic fraction: 1, β-carotene; 2, β-cryptoxanthin; 3, zeaxanthin; 4, antheraxanthin; 5, violaxanthin; 6, cucurbitaxanthin A. Red fraction: 7, capsanthin; 8, 5,6-epoxide-capsanthin; 9, capsorubin.*

Some remarkable characteristics of the carotenoid profile of *capsicum* genus include (i) the presence of provitamin A carotenoids (β-carotene and its isomeric forms, and β-cryptoxanthin), making *capsicum* products a good source of provitamin A (*6*), (ii) xanthophylls being esterified with saturated and unsaturated fatty acids (*3*, *7–9*), and the antioxidant capacity carotenoids could display in the food where they are added or in human body, once these pigments are ingested. The antioxidant properties are also indicative that these pigments will be easily degraded under oxidative conditions, such as increased light and temperature, presence of free radicals, and reactive oxygen, through which carotenoids have been demonstrated to be effective quenchers (*10*). The interaction between carotenoids and prooxidants takes place mainly through the polyene chain and any reaction or excess of energy of the environment affecting that chain will modify coloring properties of the pigment. Consequently, coloring capacity is diminished through oxidative processes and thus economic value of the product will decrease. This correlation is considered during processing of red pepper fruits for paprika, as the unit operations applied may considerably increase the progress of oxidative reactions affecting carotenoid concentration of fruits and their later stability. Hence, the suitability of the type of processing and conditions applied will be assessed in terms of minimal degradation of the carotenoid content and its maximal stability during the storage of the final product.

## Nor-Carotenoids Are Degradation Products Formed Through Thermal Stress

Processing of *capsicum* fruits for production of paprika and paprika oleoresins means a continuous application of thermal processes that produce several effects on the initial carotenoid content of the fruits. In the case of processing of dry fruits for paprika oleoresin production, an extraction step with organic solvent (hexane, methylene chloride, and ethyl acetate) is applied with a subsequent evaporation step in order to remove the solvent. In this case trends in food legislation attempts to diminish levels of residual solvents as new data about their toxicity and no observed adverse effect level values are revised and established. Additionally, food industry is required to modify those processing conditions to reduce contaminants levels when it is technically possible. Therefore, a reduction in solvent residue levels in food additives imply to change processing conditions in the desolventizing step increasing the time-temperature regime, and consequently further changes in overall characteristics of end product may occur, regarding carotenoid content and the appearance of degradation products. Isomerization, oxidation, and breakdown are the main reaction pathways described for carotenoids, and products resulting from those routes are generally a mixture of epoxides and apocarotenals. However, there is another pathway through which thermally induced reactions produce a different set of degradation products. When oleoresin samples have been subjected to a heat-induced degradation process new compounds with UV-vis. spectra similar to carotenoids are detected. They are relatively the major contributors to the carotenoid pattern when the heating time increases as can be observed in Figure 2A that shows a chromatogram corresponding to the analysis of a de-esterified carotenoid extract from a paprika oleoresin. This group was isolated and applied to positive ion APCI-HPLC-MS to determine their protonated molecular ion. From the nine compounds observed (Figure 2B) one set showed *m/z* 495 for the protonated molecular ion, corresponding to a $C_{32}H_{47}O_4$ composition, while another set showed the protonated molecular ion at *m/z* 479, corresponding to a $C_{32}H_{47}O_3$ composition. From the first set of compounds, two of them (**II** and **IV**) were isolated by preparative HPLC and analysed by EI-MS. Figure 3 shows the EI-MS spectra of compound **II** (Figure 3A) and **IV** (Figure 3B). The presence of fragments at *m/z* 476 and the peak at *m/z* 339 suggest the existence of hydroxyl and keto groups, respectively (Figure 3A, 3B).

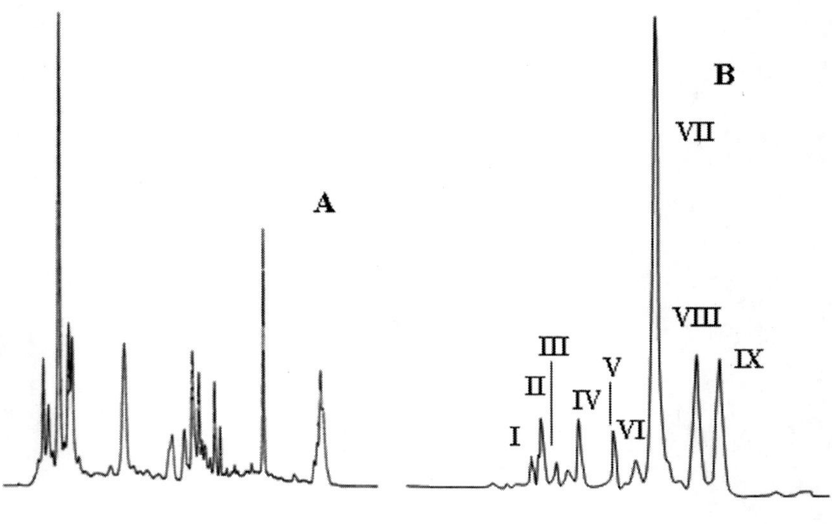


*Figure 2. HPLC traces corresponding to the analysis of a de-esterified extract from a paprika oleoresin subjected to thermal stress (A) and analysis of the isolated degradation products as described in reference (10) (B). Reproduced with permission from reference (11). Copyright (2005) (American Chemical Society).*

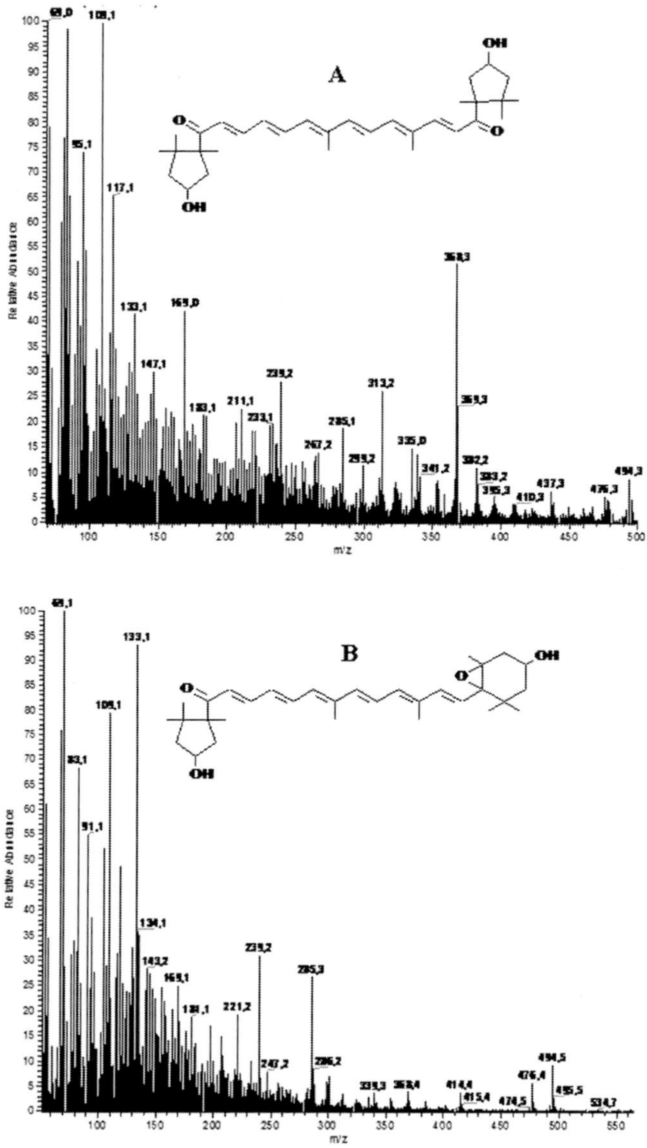


*Figure 3. EI-MS analyses of compounds II (A) and IV (B) including tentative structural assignment. Reproduced with permission from reference (11). Copyright (2005) (American Chemical Society).*

The presence of an ion at *m/z* 368 could imply the elimination of a five-membered ring. Considering the fragmentation pattern we propose the structure for compounds **II** and **IV** as octanor-capsorubin and octanor-5,6-epoxide-capsanthin (*11*). Both compounds have the same elemental composition but one of the κ rings of compound **II** is changed to a 3-hydroxy-5,6-epoxide β ring in compound **IV**. The basis for that structural differentiation is the chromatographic behaviour and the difference in the intensity of the ion at *m/z* 368 corresponding to a rearrangement, with proton transfer, from the two five-membered rings in compound **II** or one in the compound **IV**. Additionally, the EI-MS spectrum of compound **IV** denotes the presence of the peak at *m/z* 414, characteristic for epoxy carotenoids. From the second set of compounds at *m/z* 479, the major degradation product formed during thermal stress of paprika oleoresins was compound **VII**. Figure 4 depicts the EI-MS spectrum of compound **VII** which shows two successive water neutral losses and a typical elimination of an in-chain 92 u fragment from carotenoids at *m/z* 461, 443, and 387, respectively. The fragment ion at *m/z* 369 should correspond to the successive elimination of one molecule of water and an in-chain loss of a 92 u fragment. Additionally, the presence of a five-membered ring is denoted by the fragment at *m/z* 109. Thus, compound **VII** is assigned as octanor-capsanthin (*11*).

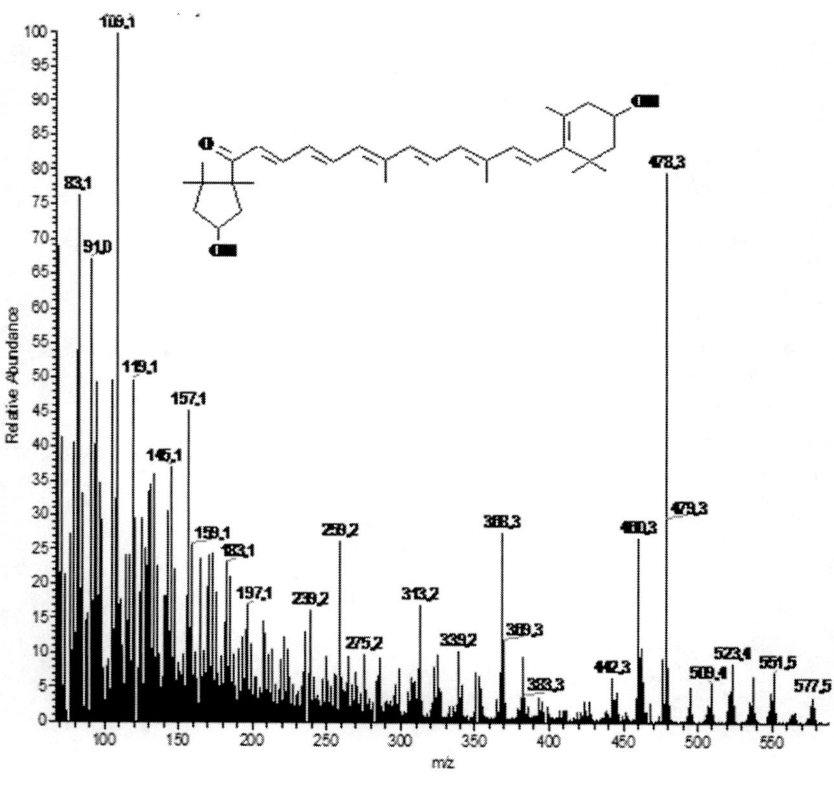


*Figure 4. EI-MS analysis of compounds VII including tentative structural assignment. Reproduced with permission from reference (11). Copyright (2005) (American Chemical Society).*

The heat-induced degradation yielded a degradation profile formed by *nor*-carotenoids, that is, the polyene skeleton has been shortened, producing a degradation product with end-groups identical to those in the parent compound, but eight carbon units have been eliminated. The pathway through which these compounds are obtained consists of the cyclization of the polyene skeleton with a subsequent elimination of an in-chain unit, yielding the nor-carotenoid and *m*-xylene. This route was proposed by Edmunds and Johnstone that identified small amounts of toluene, *m*-xylene and 2,6-dimethylnaphtalene as products of the pyrolysis of β-carotene (*12*). The conjugated double bonds of the chain can switch from *trans* to *cis* configuration and cyclisation and elimination reactions from the polyene can yield different aroma compounds and the corresponding nor-carotenoid. Consequently, the determination of the aromatic profile in paprika oleoresin rich in carotenoid pigments would have considerable importance since these compounds would be indicators of the effects of thermal impact, the extraction process and any other subsequent heat process on the product.

## Aroma Compounds Derived from Thermal Degradation of Carotenoids

The profile of volatile compounds obtained by GC-MS for paprika oleoresin is complex, with only 4 compounds reaching values exceeding 5% of the total volatile fraction (*13*). Table I lists the compounds derived from the thermal degradation of carotenoids. Two groups of compounds are distinguished: cyclic olefins and linear ketones.

**Table I. Thermodegraded Carotenoid-Derived Compounds Found in Paprika (*Capsicum annuum* L.) Oleoresin**

| *compound* | % [a] |
|---|---|
| toluene (C)[b] | 1.083 |
| *m*-xylene (C) | 7.758 |
| methylbenzaldehyde[c] (C) | 4.778 |
| 6-methyl-3,5-heptadien-2-one (L) | 5.438 |
| 6-methyl-5-hepten-2-one (L) | 0.381 |
| ethanone, 1-(methylphenyl) (C) | 6.455 |
| β-ionone (C) | 3.498 |

[a] Data are expressed as percentage values based on total peak area. [b] C means cyclic olefin, L means linear ketone. [c] Sum of different methylbenzaldehydes found in the volatile profile.

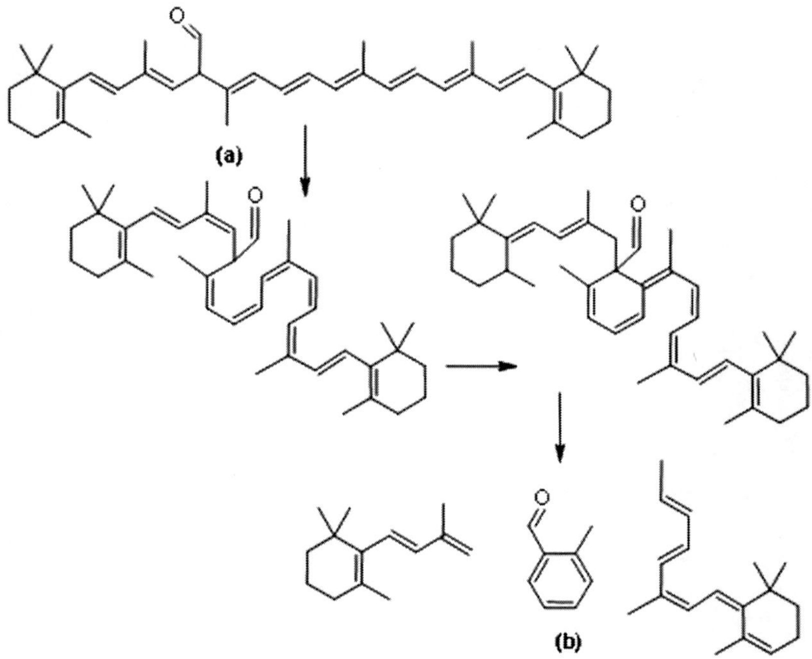


*Figure 5. Scheme of the proposed reaction for the formation of 2-methylbenzaldehyde (b) from the precursor 12-formyl-11-nor-β-carotene (a) through cyclisation and subsequent heterolysis.*

In the first group, *m*-xylene and toluene stand out, solvents that are not used for extraction of the carotenoid fraction from paprika. In view of the results *m*-xylene is one of the predominant compounds in the profile of paprika oleoresin, reaching percentages of 7.76 %; whereas toluene is present in lower percentages, (approx. 1 %). The presence of toluene and *m*-xylene, confirms the mechanism of reaction through which nor-carotenoids are generated by thermal oxidation. Formation of other volatile products follows the same reaction pathway (intramolecular cyclisation and elimination) but from structural rearrangement of an oxidized carotenoid. Thus, 2-methylbenzaldehyde and ethanone, 1-methylphenyl are produced from a previously oxidized carotenoid and the subsequent reorganization and fragmentation of the intermediate. The schematic reaction mechanism for generation of these products is shown in Figures 5 and 6, respectively. The precursors 12-formyl-11-*nor*-carotenoid (a; Figure 5) and 19-oxomethyl-10-*nor*-β-carotene (c; Figure 6) come from the peroxyl radical-mediated oxidation of β-carotene (*14*). Both reaction mechanisms and intermediates are based on the pathway proposed by Edmunds and Johnstone

(*12*). β-Ionone is detected in the profile of volatiles of the paprika oleoresin, representing 3.5 % of the total (see Table I). The production of this *apo*-carotenoid by heat-induced decomposition has been reported previously as the result of the cleavage of the double bond at position 9, 10. The profile of volatiles also includes several products (such as ketones 6-methyl-3,5-heptadien-2-one and 6-methyl-5-hepten-2-one) which are generated when oxidation of the polyene chain is the promoted reaction process.

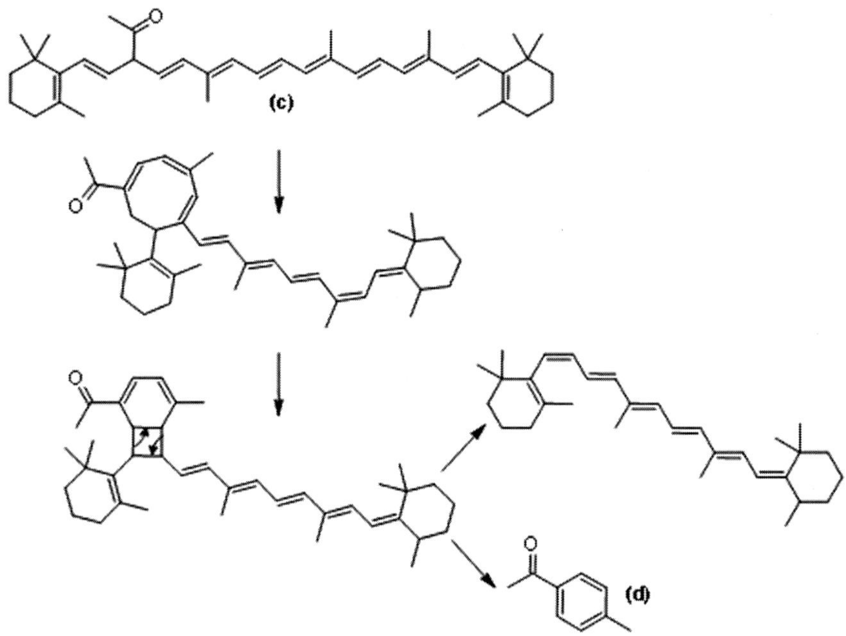


*Figure 6. Scheme of the reaction proposed for the formation of ethanone-1,(methylphenyl) (d) from the precursor 19-oxomethyl-10-nor-β-carotene (c) through intramolecular cyclisation and subsequent in-chain loss unit.*

The use of vigorous processing conditions to obtain paprika oleoresin rich in color, and the subsequent operations to remove solvent residues, facilitate the reaction process discussed above. Intramolecular cyclisation stands out as the main reaction of structural transformation of carotenoids in paprika oleoresin. *Trans-cis* isomerization reaction is the previous step of the intramolecular cyclation, so it must be taken into account that the observed degradation product patterns could be a consequence of the different tendency that each carotenoid shows to form poly-cis isomers. This process as well as those that take place subsequently (elimination within the chain or heterolytic fragmentation), are promoted by a higher temperature. Obviously, the application of a more vigorous temperature-time regime, aimed at increasing the extraction yield of pigments from the raw material, or to minimize residual solvents, would result in a greater accumulation of degradation products. The use of a milder processing or an extraction technique with minimal thermal impact (such as extraction with supercritical fluid) is more likely to produce a lower oxidation index. Thus, the presence of aromatic compounds generated by thermal degradation could be used as a quality marker of the processing, and even distinguish the type of extraction technique employed. Volatiles originated from thermal degradation of carotenoids have found application for detection of microbial contamination in canned fruits (*15*), and as biomarkers in rumiant tissues for dietary discrimination among animals feed with pasture diet and concentrate (*16*, *17*).

## References


1. Davies, B. H.; Matthews, S.; Kirk, J. T. O. *Phytochemistry* **1970**, *9*, 797–805.
2. Mínguez-Mosquera, M. I.; Garrido-Fernández, J.; Pereda-Marín, J. *Grasas Aceites (Sevilla, Spain)* **1984**, *35*, 4–10.
3. Mínguez-Mosquera, M. I.; Hornero-Méndez, D. *J. Agric. Food Chem.* **1994**, *42*, 38–44.
4. Jarén-Galán, M.; Mínguez-Mosquera, M. I. *J. Agric. Food Chem.* **1999**, *47*, 4379–4383.
5. Jarén-Galán, M.; Pérez-Gálvez, A.; Mínguez-Mosquera, M. I. *J. Agric. Food Chem.* **1999**, *47*, 945–951.
6. Lachance, P. *Clin. Nutr.* **1988**, *7*, 118–122.
7. Camara, B.; Monéger, R. *Phytochemistry* **1978**, *17*, 91–93.
8. Biacs, P. A.; Daood, H. G.; Pavisa, A.; Hajdu, F. *J. Agric. Food Chem.* **1989**, *37*, 350–353.
9. Mínguez-Mosquera, M. I.; Hornero-Méndez, D. *J. Agric. Food Chem.* **1994**, *42*, 640–644.
10. Foote, C. S.; Denny, R. W. *J. Am. Chem. Soc.* **1968**, *90*, 6233–6235.
11. Pérez-Gálvez, A.; Ríos, J. J.; Mínguez-Mosquera, M. I. *J. Agric. Food Chem.* **2005**, *53*, 4820–4826.
12. Edmunds, F. S.; Johnstone, R. A. W. *J. Chem. Soc.* **1965**, *46*, 2892–2897.
13. Ríos, J. J.; Fernández-García, E.; Mínguez-Mosquera, M. I.; Pérez-Gálvez, A. *Food Chem.* **2008**, *106*, 1145–1153.
14. Yamauchi, R.; Miyake, N.; Inoue, H.; Kato, K. *J. Agric. Food Chem.* **1993**, *41*, 708–713.
15. Bianchi, F.; Careri, M.; Mangia, A.; Mattarozzi, M.; Musci, M.; Concina, I.; Falasconi, M.; Gobbi, E.; Pardo, M.; Sberveglieri, G. *Talanta* **2009**, *77*, 962–970.
16. Guilhem, S.; Jeremy, R.; Erwan, E. *Food Chem.* **2010**, *118*, 418–425.
17. Valentina Vasta, V.; Luciano, G.; Dimauro, C.; Röhrle, F.; Priolo, A.; Monahan, F. J.; Moloney, A. P. *Meat Sci.* **2011**, *87*, 282–289.

## Chapter 18

# Egg Yolk Carotenoids: Composition, Analysis, and Effects of Processing on Their Stability

**Chamila Nimalaratne,[1] Jianping Wu,[1] and Andreas Schieber*,[1,2]**

**[1]Department of Agricultural, Food and Nutritional Science, University of Alberta, 4-10 Agriculture/Forestry Centre, Edmonton, AB T6G 2P5, Canada**
**[2]Department of Nutritional and Food Sciences, University of Bonn, Römerstrasse 164, D-53117 Bonn, Germany**
***E-mail: schieber@uni-bonn.de.**

Carotenoids present in egg yolk have been demonstrated to be highly bioavailable, which makes eggs an interesting carrier of these biologically active pigments. The profile of carotenoids can be manipulated through the diet of laying hens. Eggs are usually cooked prior to consumption, which may lead to degradation and isomerization of carotenoids. This review summarizes selected studies on the composition of egg yolk carotenoids, their changes upon heating, and also provides a brief overview of common methods for carotenoid analysis.

Eggs constitute an important source of macronutrients and micronutrients such as proteins, lipids, vitamins and minerals. The protein fraction not only delivers essential amino acids but part of the proteins and derived peptides also show biological activities, for example, antimicrobial, immunomodulating, antihypertensive, and antioxidant properties (*1*, *2*). The biological activities of ovotransferrin, which accounts for approximately 12-13% of total egg white protein, have very recently been summarized (*3*). Interestingly, free aromatic amino acids, especially tryptophan and tyrosine, contribute to the antioxidant properties of egg yolk (*4*).

In addition to providing both basic nutrients and endogenous bioactive components, eggs have also been used as a vehicle of nutraceutical compounds. The best-known example are probably omega-3 fatty acids, which are incorporated into eggs by feeding fish oil, flax seed, algae, or other ingredients to laying hens (*5*). In this respect, eggs are more suitable compared to foods derived from ruminants (milk, fat) because extensive microbial biohydrogenation in the rumen leads to significant conversion of the polyunsaturated fatty acids (*6*).

Carotenoids represent another group of lipid soluble bioactive compounds that are present in eggs and that give egg yolk the yellow, orange or golden color that is appreciated by consumers. These pigments are usually classified into carotenes, which are pure hydrocarbons, and xanthophylls, which contain one or more oxygen atoms. Over the past decades, carotenoids have attracted intense interest by food scientists and researchers from other disciplines because of their health promoting properties (*7*). Their occurrence, structures, chemistry and analysis, biological activities, bioavailability, and stability during processing of various foods have extensively been reviewed (*8–14*) and will not be covered here. Instead, the objective of this contribution is to provide an overview of egg yolk carotenoids in relation to the composition, methods of extraction and analysis, and effects of processing and storage on their stability.

## Carotenoids in Egg Yolks

The profile of carotenoids in egg yolk is highly dependent on the hens´ diet. In other words, the type and amount of carotenoids in yolk can be manipulated through poultry feed handling. Similarly, different rearing systems produce eggs with distinct yolk carotenoid composition because of the differences in feed utilization (*15*). The concept of 'designer eggs' is used for the eggs enriched with certain nutrients such as vitamin E, omega-3 fatty acids, selenium, and carotenoids through feed manipulation. With the altered nutrient composition, designer eggs are also considered a new type of functional food (*16*).

In the European Union, eight carotenoids are approved as additives in poultry feed. These are capsanthin (C40), β-cryptoxanthin (C40), lutein (C40), zeaxanthin (C40), β-apo-8′-carotenal (C30), β-apo-8′-carotenoic acid ethyl ester (C30), canthaxanthin (C40), and citranaxanthin (C33) (*17*). In Canada, only β-apo-8′-carotenoic acid ethyl ester, lutein and canthaxanthin are permitted according to the Canadian Feeds Regulation issued in 1983. Generally, lutein, capsanthin, zeaxanthin, and β-cryptoxanthin are obtained from plant sources while others are produced through chemical synthesis. They are added in crystalline form and are subject to maximum regulatory limits. In the EU, the limit is 80 mg of synthetic pigment per kg of finished feed except for canthaxanthin, for which of limit of 8 mg/ kg feed has been set. In Canada, the limit is 30 mg/ kg feed for canthaxanthin and β-apo-8′-carotenoic acid ethyl ester. The United States Food and Drug Administration sets limits for canthaxanthin as ≤ 30 mg/lb of solid or semisolid food or per pint of liquid food for general use and ≤ 4.41 mg per kg for broiler chicken feed. In addition, dried algal meal, corn endosperm oil and Aztec marigold meal and extract are the only compounds permitted as poultry color additives.

As a result of the above mentioned differences in the legal status, the profile of egg yolk carotenoids may vary significantly from country to country and also with the rearing system. In most cases, lutein obtained from natural sources such as marigold (*Tagetes erecta*) or alfalfa (*Medicago sativa*) extracts is the predominant carotenoid in egg yolks. Other sources of xanthophylls used in poultry feeding are corn (*Zea mays*) and red pepper (*Capsicum annuum*). Canthaxanthin, β-apo-8′-carotenal and β-apo-8′-carotenoic acid ethyl ester are chemically synthesized (*18*). Since hens are efficiently converting β-carotene to vitamin A, only minor quantities of this otherwise widespread carotenoid are found in egg yolks.

Vegetables and fruits such as carrots, spinach, corn, tomatoes, pumpkins, mangoes, and papayas are considered the most important dietary sources of carotenoids. However, the bioavailability of carotenoids from plants may be poor and dependent, among others, on the physical state of the pigments, processing, and the presence of lipids (*19*). In contrast, xanthophylls from egg yolks have been demonstrated to have a higher bioavailability (*20*, *21*). Since moderate egg consumption is no longer associated with an increased risk of developing coronary heart disease (*22*), egg yolk can be considered an ideal carrier of biologically active carotenoids for human consumption.

## Effects of Processing and Storage on Egg Yolk Xanthophylls

The polyene structure renders carotenoids susceptible to isomerization and degradation caused by conditions typically applied during processing of foods. Especially upon exposure to high temperatures and UV light, and in the presence of oxygen, certain enzymes such as mono- and dioxygenases, and redox active metal ions, all-*trans* carotenoids may be converted to their *cis*-isomers (*14*) or oxidized and cleaved to apo-carotenoids (*23*, *24*). On one side, these conversions may be associated with a partial or complete loss of bioactivity, on the other side, the degradation products may show entirely different biological activities (*25*). Heat treatment has been demonstrated to cause the formation mainly of 13-*cis* isomers, whereas exposure to light leads to the 9-*cis* isomers of carotenoids. The presence of lipids may facilitate the formation of *cis*-isomers since carotenoids initially present in crystalline form may be dissolved, which makes them more prone to isomerization (*26*).

Eggs are mostly eaten in cooked form. The most common ways of domestic cooking are boiling, frying, scrambling, and microwaving. Industrial preparation of egg products such as yolk powder and liquid egg yolk may involve freeze drying, spray drying and pasteurization. In addition, several non-thermal techniques, e.g., gamma irradiation and high pressure treatment have been reported as an alternative to high temperature processes in an effort to minimize damage to heat sensitive components. Surai et al. (*27*) found that lutein in ′designer eggs′ is not degraded during boiling. In contrast, a considerable decrease (10-20%) in egg yolk xanthophylls during boiling was observed by other authors (*15*), who also reported that lutein was most susceptible, whereas β-apo-8′-carotenoic acid ethyl ester proved to be most stable. In view of the above mentioned properties of carotenoids, and considering that eggs constitute a matrix rich in lipids, one would expect that significant isomerization of egg yolk xanthophylls will occur during processing. Surprisingly, our recent investigations into the stability of egg yolk xanthophylls during domestic cooking revealed that the qualitative profile of carotenoid stereoisomers does not change upon heating. Boiling, frying and microwave heating led to a significant decrease especially in all-*trans* lutein and a concomitant slight increase in the 13-*cis* isomers of lutein and zeaxanthin, which however was less pronounced than expected. Total losses of xanthophylls ranged from 6-18%.

Storage for 6 months at -18 °C and 20 °C resulted in significant losses of xanthophylls in both pasteurized and non-pasteurized freeze dried egg yolk. The decrease in carotenoids occurred mainly within the first 4 weeks of storage. Canthaxanthin and β-apo-8′-carotenoic acid ethyl ester were more stable than lutein and zeaxanthin. It was also observed that temperature had no influence on the content of xanthophylls in freeze dried egg yolks during storage for 6 months in darkness (*28*).

Among the non-thermal techniques, ionizing radiation has been used to eliminate *Salmonella* in liquid and frozen egg products and was shown to decrease total carotenoids (*29*, *30*). Follow-up studies demonstrated that carotenoid degradation in egg yolks is caused mainly by the action of irradiation induced lipid peroxides. Carotenoids were more stable when the treatment was carried out *in vacuo* and products stored in an oxygen free environment (*31*, *32*). A significant decrease in total carotenoids was observed after gamma irradiation followed by cold storage at 4 °C of liquid egg yolk; cold storage led to carotenoid losses also in non-irradiated samples (*30*). Lipid oxidation and degradation of carotenoids was observed also when liquid egg samples were irradiated at a dose of 3 kGy. In contrast to irradiation treatment, the use of high hydrostatic pressure (5 MPa for 5 min) did not lead to changes in the carotenoid content of eggs (*33*). Vacuum packaging and the presence of antioxidants such as vitamin E and butylated hydroxytoluene was shown to reduce color changes in egg yolk powder during radiation, which is indicative of a protective effect toward xanthophylls (*34*).

## Analysis of Egg Yolk Carotenoids

Eggs contain approximately 75% water, 12% lipids, and 12% proteins, as well as some minor quantities of carbohydrates and minerals (*1*). Since the lipid fraction is almost exclusively found in the yolk, the extraction of carotenoids from this complex matrix is challenging, especially since they are susceptible to heat and light and may be degraded in the presence of lipid peroxides. Furthermore, xanthophylls may be present both in free form and as fatty acid esters, which may affect their polarity and hence their solubility.

Most of the procedures reported for the analysis of egg yolk carotenoids rely on their extraction with organic solvents followed by a purification step to remove co-extracted lipids that would otherwise interfere with the subsequent analysis (*35*). Alkaline saponification not only facilitates the removal of triglycerides but also leads to a less complex profile of carotenoids by cleavage of xanthophyll esters. However, information about the acylating moiety is lost and alkali labile carotenoids may suffer from partial degradation, leading to an underestimation of their concentration. In early studies, mixtures of petroleum ether and alcohol were used for the extraction of total carotenoids, separating the xanthophylls into the alcohol phase and the non-polar carotenes into the ether phase (*36*, *37*), whereas later, acetone was the solvent of choice (*38*, *39*). An excellent recovery of >99% was found when a ternary solvent system consisting of light petroleum ether, ethyl acetate and methanol (1:1:1, v/v/v) was used for the extraction of xanthophylls from egg yolk (*15*). Combined ultrasound treatment and solvent extraction was proven to be more effective in extracting lutein from egg yolk (*40*). Ultrasound assisted extraction was also used in more recent studies on the effects of processing and storage on carotenoid contents in egg yolk (*28*, *41*).

Nowadays, HPLC with diode array and mass spectrometric detection (LC-DAD-MS) is the method of choice for the qualitative characterization and quantitative determination of carotenoids in various matrices (*13*), and egg yolk carotenoids are obviously not an exception. In many cases, LC-DAD-MS provides maximum information about the nature of the carotenoid, the acylating moiety in xanthophyll esters, and even about the stereochemistry, i.e., *cis-trans* isomers. Normal-phase and $C_{18}$ stationary phase used in earlier studies have largely been replaced by $C_{30}$ columns, which allow for a superior resolution of carotenoid geometric and positional isomers (*13*). Ultra high-performance liquid chromatography (UHPLC) is a relatively new technique, which is characterized by an increase in resolution, sensitivity, and speed of analysis. Consequently, it allows for a higher sample throughput and is more cost effective (*42*). Although a number of studies for the fast determination of carotenoids have been reported (*13*), it appears that the potential of UHPLC has not yet fully been exploited. Initial investigations carried out in our laboratory demonstrated that UHPLC is capable of separating egg yolk carotenoids, including *cis*-isomers of lutein and zeaxanthin, within less than 10 min, whereas approximately 80 min were required using conventional HPLC.

## Conclusions

Eggs are an important food commodity and a source of various biologically active components. Xanthophylls from egg yolks have been demonstrated to be highly bioavailable, which renders eggs an interesting carrier of carotenoids in the human diet. Their profile can be manipulated through administration of carotenoids in the diet of laying hens. The observation that carotenoids are relatively stable toward isomerization during processing merits further attention, especially with respect to the stabilizing principles. From an analytical point of view it seems that there is still a need to develop methods for a faster extraction and determination of carotenoids from egg yolks. So far, investigations into the formation of carotenoid cleavage products during processing, in particular during frying and scrambling of eggs, are missing.

# References


1. Kovacs-Nolan, J.; Phillips, M.; Mine, Y. *J. Agric. Food Chem.* **2005**, *53*, 8421–8431.
2. Mine, Y. *Curr. Pharm. Des.* **2007**, *13*, 875–884.
3. Wu, J.; Acero-Lopez, A. *Food Res. Int.* **2012**, *46*, 480–488.
4. Nimalaratne, C.; Lopes-Lutz, D.; Schieber, A.; Wu, J. *Food Chem.* **2011**, *129*, 155–162.
5. Gonzalez-Esquerra, R.; Leeson, S. *Can. J. Anim. Sci.* **2001**, *81*, 295–306.
6. Scollan, N. D.; Dhanoa, M. S.; Choi, N. J.; Maeng, W. J.; Enser, M.; Wood, J. D. *J. Agric. Sci.* **2001**, *136*, 345–356.
7. Rao, A. V.; Rao, L. G. *Pharmacol. Res.* **2007**, *55*, 207–216.
8. Maiani, G.; Caston, M. J. P.; Catasta, G.; Toti, E.; Cambrodon, I. G.; Bysted, A.; Granado-Lorencio, F.; Olmedilla-Alonso, B.; Knuthsen, P.; Valoti, M.; Böhm, V.; Mayer-Miebach, E.; Behsnilian, D.; Schlemmer, U. *Mol. Nutr. Food Res.* **2009**, *53*, S194–S218.
9. Fernández-García, E.; Carvajal-Lérida, I.; Jarén-Galán, M.; Garrido-Fernández, J.; Pérez-Gálvez, A.; Hornero-Méndez, D. *Food Res. Int.* **2012**, *46*, 438–451.
10. Britton, G. *FASEB J.* **1995**, *9*, 1551–1558.
11. Namitha, K. K.; Negi, P. S. *Crit. Rev. Food Sci. Nutr.* **2010**, *50*, 728–761.
12. Boon, C. S.; McClements, D. J.; Weiss, J.; Decker, E. A. *Crit. Rev. Food Sci. Nutr.* **2010**, *50*, 515–533.
13. Rivera, S. M.; Canela-Garayoa, R. *J. Chromatogr., A* **2012**, *1224*, 1–10.
14. Schieber, A.; Carle, R. *Trends Food Sci. Technol.* **2005**, *16*, 416–422.
15. Schlatterer, J.; Breithaupt, D. *J. Agric. Food Chem.* **2006**, *54*, 2267–2273.
16. Surai, P. F.; Sparks, N. H. C. *Trends Food Sci. Technol.* **2001**, *12*, 7–16.
17. List of Authorized Additives in Feedingstuff. *Council Directive 70/524/EEC.*
18. Breithaupt, D. E. *Trends Food Sci. Technol.* **2007**, *18*, 501–506.
19. Zaripeh, S.; Erdman, J. W. *J. Nutr.* **2002**, *132*, 531S–534S.
20. Handelman, G. J.; Nightingale, Z. D.; Lichtenstein, A. H.; Schaefer, E. J.; Blumberg, J. B. *Am. J. Clin. Nutr.* **1999**, *70*, 247–251.
21. Chung, H.; Rasmussen, H. M.; Johnson, E. J. *J. Nutr.* **2004**, *134*, 1887–1893.
22. Fernandez, M. L. *Curr. Opin. Clin. Nutr. Metab. Care* **2006**, *9*, 8–12.
23. Fleischmann, P.; Zorn, H. In *Carotenoids Volume 4: Natural Functions*; Britton, G., Liaaen-Jensen, S., Pfander, H., Eds.; Birkhäuser: Basel, Switzerland, 2008; pp 341−366.
24. Carail, M.; Caris-Veyrat, C. *Pure Appl. Chem.* **2006**, *78*, 1493–1504.
25. Wang, X. D. In *Carotenoids in Health and Disease*; Krinsky, N. I., Mayne, S. T., Sies, H., Eds.; Marcel Dekker: New York, 2004; pp 313−335.
26. Marx, M.; Stuparić, M.; Schieber, A.; Carle, R. *Food Chem.* **2003**, *83*, 609–618.
27. Surai, P. F.; MacPherson, A.; Speake, B. K.; Sparks, N. H. C. *Eur. J. Clin. Nutr.* **2000**, *54*, 298–305.
28. Wenzel, M.; Seuss-Baum, I.; Schlich, E. *Food Chem.* **2011**, *124*, 1343–1348.
29. Katusin-Razem, B.; Razem, D.; Matic, S.; Mihokovic, V.; Kostromin-Soos, N.; Milanovic, N. *J. Food Prot.* **1989**, *52*, 781–786.

30. Badr, H. M. *Food Chem.* **2006**, *97*, 285–293.
31. Katusin-Razem, B.; Mihaljevic, B.; Razem, D. *J. Agric. Food Chem.* **1992**, *40*, 662–668.
32. Katusin-Razem, B.; Mihaljevic, B.; Razem, D. *J. Agric. Food Chem.* **1992**, *40*, 1948–1952.
33. Andrassy, E.; Farkas, J.; Seregély, Z.; Dalmadi, I.; Tuboly, E.; Lebovics, V. *Acta Aliment.* **2006**, *35*, 305–318.
34. Du, M.; Ahn, D. *J. Food Sci.* **2000**, *65*, 625–629.
35. Schiedt, K.; Liaaen-Jensen, S. In *Carotenoids Volume 1A:Isolation and Analysis*; Britton, G., Liaaen-Jensen, S., Pfander, H., Eds.; Birkhäuser: Basel, Switzerland, 1995; pp 81−108.
36. Palmer, L. S. *J. Biol. Chem.* **1915**, *23*, 261–279.
37. Hughes, J.; Payne, L. *Poult. Sci.* **1937**, *16*, 135–138.
38. Marusich, W.; De Ritter, E.; Bauernfeind, J. *Poult. Sci.* **1960**, *39*, 1338.
39. Smith, I.; Perdue, H. *Poult. Sci.* **1966**, *45*, 577–581.
40. Yue, X.; Xu, Z.; Prinyawiwatkul, W.; King, J. M. *J. Food Sci.* **2006**, *71*, C239–C241.
41. Wenzel, M.; Seuss-Baum, I.; Schlich, E. *J. Agric. Food Chem.* **2010**, *58*, 1726–1731.
42. Swartz, M. E. *J. Liq. Chromatogr. Relat. Technol.* **2005**, *28*, 1253–1263.

# Editors' Biographies

## Peter Winterhalter

Peter Winterhalter is head of the Institute of Food Chemistry at the Technische Universität Braunschweig (TUBS), Germany. He graduated from the University of Karlsruhe and obtained his Ph.D. from Würzburg University in 1988. In 1989 he worked as a postdoctoral fellow at the Australian Wine Research Institute, Adelaide, before returning to Würzburg University as a Research Fellow. In 1995, he was appointed as Professor of Food Chemistry at the University of Erlangen-Nürnberg before finally moving to TUBS where he was dean of the Faculty of Chemistry and Pharmacy. His research activities focus on natural products/food chemistry and include aroma precursors, countercurrent chromatography, wine research, food authentication, and natural pigments.

## Susan E. Ebeler

Susan E. Ebeler is a professor in the Department of Viticulture and Enology at the University of California, Davis. Her research is focused on development and application of analytical chemistry techniques to study flavor chemistry and the physico-chemical interactions of flavors with nonvolatile wine components. She also studies the chemical mechanisms for the observed health effects of food and wine components. At UC Davis, she teaches undergraduate and graduate classes on the analysis of grapes and wines as well as the flavor chemistry of foods and beverages. She is also co-director of the UC Davis Food Safety and Measurement Facility which seeks to advance the chemical understanding of foods and beverages for safety, authentication, and quality.

# Indexes

# Author Index

# Subject Index

## A



## B



## C

# L



# M



# N



# O

## P



## R